Digital Assembly Coordination and Quality Controlling Technology for Aeronautical Thin-walled Structures

航空薄壁结构的数字化装配协调与质量控制技术

Feiyan Guo   Zhongqi Wang   Shaozhuo Li
郭飞燕 王仲奇 李少卓 著

化学工业出版社
·北京·

内容简介

装配中的不协调，是飞行器生产中的主要矛盾和中心问题，解决该问题是保证飞行器制造/装配质量的前提。本书针对新一代飞行器的高性能设计、制造和装配要求，系统介绍了飞行器数字化装配协调技术和装配质量控制技术的科学理论与工程方法，主要包括基于数字协调模型的飞行器装配工作方法、产品关键协调特征的量化识别、柔性装配工装系统定位精度的精准保障、装配误差的传递与协调误差尺寸链的构建、装配质量的闭环控制与其可靠性提升等内容。一方面，本书结合最新研究进展，以具体的装配协调质量控制需求为出发点，对装配工艺与协调技术进行系统的解析；另一方面，以飞行器制造的工艺过程为核心，强化理论研究与工程现场的综合应用，帮助读者更好地理解飞行器装配协调技术的理论方法。

本书内容丰富，可供航空航天科学与技术学科的教师和学生以及装配质量相关的科研人员与工程技术人员参考使用。

Inconsistency in assembly is the main contradiction and central issue in aeronautical production, and resolving this inconsistency is a prerequisite for ensuring the quality of aeronautical manufacturing and assembly. Focusing on the high performance requirements for design and manufacturing and assembly, this book systematically introduces the scientific theory and engineering methods of full digital assembly coordination technology and assembly quality control technology. Specific content includes aircraft assembly work methods based on digital coordination models, quantitative identification of key coordination features for products, precise guarantees of positioning accuracy for flexible assembly tooling systems, transmission of assembly errors and construction of coordination error chains, closed-loop control of assembly quality and improvement of its reliability. On the one hand, based on the latest research progress and taking the specific assembly coordination quality control requirements as the starting point, a systematic review on assembly processes and coordination technologies is conducted. On the other hand, focusing on the manufacturing process of aeronautical products and strengthening the comprehensive application of theoretical research and practical engineering, readers will gain an understanding of the theoretical methods of aeronautical assembly coordination technology.

For the assembly process design and assembly coordination assurance, this book provides a rich and in-depth introduction. It can be used as a reference for teachers and students in aerospace science and technology majors, as well as the researchers and engineering technicians related to assembly quality.

图书在版编目（CIP）数据

航空薄壁结构的数字化装配协调与质量控制技术 = Digital Assembly Coordination and Quality Controlling Technology for Aeronautical Thin-walled Structures：英文 / 郭飞燕，王仲奇，李少卓著. -- 北京：化学工业出版社，2025. 3. -- ISBN 978-7-122-47239-7

Ⅰ. V214.4

中国国家版本馆 CIP 数据核字第 2025KU4764 号

责任编辑：严春晖　张海丽　　　装帧设计：刘丽华
责任校对：张茜越

出版发行：化学工业出版社
　　　　　（北京市东城区青年湖南街 13 号　邮政编码 100011）
印　　装：中煤（北京）印务有限公司
710mm×1000mm　1/16　印张 14¾　彩插 5　字数 305 千字
2025 年 3 月北京第 1 版第 1 次印刷

购书咨询：010-64518888　　　　　售后服务：010-64518899
网　　址：http://www.cip.com.cn
凡购买本书，如有缺损质量问题，本社销售中心负责调换。

定　　价：138.00 元　　　　　　　　版权所有　违者必究

PREFACE

For the new generation aeronautical products, their manufacturing and assembly quality affects the service performance even more directly. In assembly work, key quality indexes, such as profile flush, gap, the convex-concave value of the fastener joint's head, have been enhanced nearly an order of magnitude compared to previous aviation type, namely, the assembly quality requirements have reached the sub-millimeter level. And the assembly coordination accuracy is generally even tighter than individual parts' manufacturing errors. For example, ① the assembly steps between different skin panels along the flight direction are not permitted for modern aircraft, ② the gaps between different skin sections are smaller than the forming/manufacturing accuracy of the sheet parts, especially for the carbon fiber reinforced polymer (CFRP) parts with relatively low manufacturing accuracy. As a result, the assurance on assembly and coordination accuracy is of great concern to the design and manufacturing departments in the modern aviation industry.

As a key technology in aircraft manufacturing, assembly coordination technology is important for improving the competitiveness and manufacturing capabilities. The main topics of this book are as follows. ① The working mode based on digital coordination model (DCM), which provides an overall guiding line for digital assembly coordination theory. ② The assembly quality analysis and control technical system, which enhances accuracy reliability and ensures assembly accuracy within the required range, and lays a precision foundation for products' service performance.

This book contains six chapters. Some content was published in the Journal of Manufacturing Systems, Composite Structures,

Robotics and Computer-Integrated Manufacturing, Advanced Engineering Informatics, International Journal of Advanced Manufacturing Technology, Assembly Automation, IEEE Access, Proceedings of the Institution of Mechanical Engineers Part B-Journal of Engineering Manufacture, and others. This book could open a door for readers to gain a deeper understanding of the assembly process and quality control technology for aviation thin-walled components. This book could also serve as a reference for teachers and students majoring in aerospace manufacturing engineering, mechanical engineering, automotive engineering, and related fields, as well as researchers and engineering technicians.

We express our special thanks for the support of the National Natural Science Foundation of China (52175450, 51805502), the National Defense Basic Scientific Research Foundation (JCKY 2023205B006), and the Aerospace Science Research Foundation (2017ZE25005). As a result of research, some content in this book may inevitably be incomplete, contain inappropriate expressions or biased opinions. We sincerely welcome feedback and corrections from experts and readers.

Contents

Chapter 1
Introduction

1.1 Background and Research Requirements for Assembly Technology

1.1.1 Background and Foundation Knowledge of Aeronautical Assembly

Aviation products and equipment are an important pillar for safeguarding national sovereignty, showing great significance to national construction and economy by defending national development interests and protecting national security. The aviation manufacturing industry has typical characteristics of high-tech and advanced manufacturing. As a result, it is a significant symbol of science and technology, economy, and industrialization level. Considering the lightweight requirement, the core load-bearing parts, such as the skeleton and skin panels, are typically comprised by numerous thin-walled structures jointed in a preset order. The typical weak rigid parts, such as skins, stringers, frames, rivets and connecting clips, are shown in Figure 1-1. Due to the complex alternating load and impact load during service period, the accuracy of structure's assembly dimensions and the state/distribution of internal interactive stress directly affect the flight performance, fatigue resistance, stress resistant and corrosion resistant capability of the whole product. Assembly process and operations act as the system integration link for these highly complex products, providing the final guarantee for assembly error control during the product development. Therefore, research on assembly quality control is of great concern to corresponding technicians[1].

For the assembly process of lightweight aviation thin-walled structures, complex process links such as product design, part/component manufacturing, assembly process planning, fixture/tooling positioning, hole-drilling and joining, as well as measurement or inspection, are typically all contained. These assembly operations and procedures accounting for more than 50% of the workload at the practical assembly site. Therefore, the assembly quality control are facing the following difficulties:

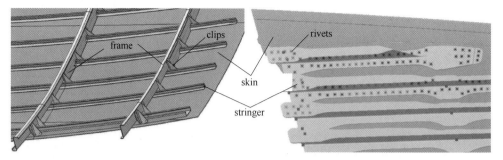

Figure 1-1   Typical thin-walled aviation assembly structures

① A large number of parts with large dimension and weak stiffness;

② High aerodynamic shape requirements, as the aviation products need smooth streamline for flight performance;

③ An extremely long assembly process with multi-stations (Figure 1-2), a tremendous amount of error links and complex error propagation path (Figure 1-3[2]);

④ A multitude of supporting and matching assembly fixture/equipment, whose accuracy is usually 3~5 times tighter than that of the product parts;

⑤ The universal tolerance and fit theory cannot guarantee the final assembly and coordination requirements, leading to the need for a complex and sound interchange and coordination principle.

As a result, current assembly theory and method that based on geometric quantity control has typical characteristics of "digital/empirical, passive repairing, and geometric shape matching", which cannot guarantee the assembly quality of complex thin-walled structures effectively, namely:

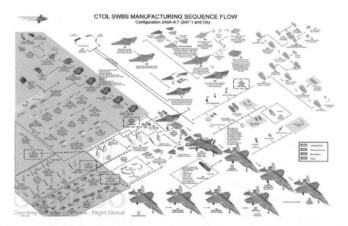

**Figure 1-2   Schematic diagram of manufacturing sequence flow with
multi-station processes for aircraft (See the color illustration)**

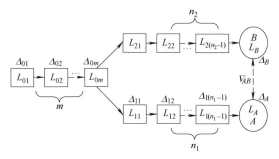

Figure 1-3   Assembly error propagation and coordination process[2]

① Emphasizing the "geometric shape control", i.e. the core is to guarantee the precision of assembly geometry size;

② The modeling and simulation for the assembly accuracy prediction is based on the ideal state of product's design model, having a weak control over the time-varying process elements on practical assembly site;

③ A large number of dedicated physical prototypes are essential to assist assembly operations, which lead to a complex assembly and verification environment and a large number of trial assemblies;

④ The assembly system lacks the function of feedback improvement, and the assembly quality assurance method mainly depends on parameter compensation and workers' manual ability;

⑤ The process parameters mostly rely on empirical design, showing a strong randomness;

⑥ The deficiency of quantitative analysis and scientific guidance are deficient results in a poor assembly error consistency across different assemblies.

The assembly performance guarantee of aviation thin-walled structures with a low rigidity has become an important solution for improving the overall development level and the core competitiveness for aviation products. From the perspective of engineering application, to overcome the above shortcomings in geometric quality control, the "Zero-Gap" concept applied on the American 5th generation aircraft F-35 represents the latest progress of profile quality control for stealth aircraft. In the integrated assembly of the monolithic skin panels, inlets, skin gaps, and hatch covers, etc., size elements and the tolerances of structure's profiles are refined and controlled. In addition, during the manufacturing of the F-35, special attention had been paid to the huge benefits generated by virtual simulation operations. A high-confidence and virtual-real fusion model is developed in JSE (joint simulation environment), namely F-35 In A Box, or FIAB. JSE is a scalable, extensible, high-fidelity, and non-proprietary virtual commissioning environment, and it can overcome the time and space constraints of physical experiments. Then a lot of testing and verification work could be completed in the laboratory in advance. According to statistics, although the United States faced severe challenges from the Epidemic in 2020, the F-35 annual production capacity still reached 123 (compared to 91 in 2018)[3]. From another perspective, in order to reduce the structure weight and pursue rapid, accurate assembly, the "2mm project" in the automotive manufacturing industry was adopted. In aviation industry, the "0.1mm project" was introduced into structure design, requiring that the difference between designed and manufactured parts not exceed 0.1mm. Only in this way can the service performance of the new generation aviation products be effectively guaranteed. And with the help of full 3D information and integrated manufacturing technology, as well as the control methods such as parts precise manufacturing, tolerance analysis, digital quantity coordination, measurement-based virtual assembly, assembly compensation simulation, and other assembly gap/step controlling operations, the aerodynamic profile accuracy requirements can be guaranteed. For the "adaptive assembly" concept, with the full use of digital measurements, only "Installation" operations could meet the assembly performance requirements, contrasting obviously with the current method of "Repair firstly, then assembly". However, it mainly orients to the practical manual operations for large component docking and assembly[4], lacking the support of basic assembly methods such as quantitative prediction and precise control of assembly accuracy and internal interactive stress.

In a full digital coordination environment, digital measurement technology is taken as the basic premise for implementing the transfer of coordination relationships. The manufacturing and assembly accuracy of product parts and tooling equipment relies more on the measured data, specifically the spatial position and its corresponding error. With these measured data from the practical assembly site, the adjustments of assembly process can be supported by evidence, and the assembly error problem can be detected and controlled in time. However, measurement data only indicates whether the coordination accuracy at a certain manufacturing stage meets the design requirements. The detailed reasons behind the inconsistency phenomenon among different assemblies (Figure 1-3) in the coordination process cannot be explained.

In order to ensure the coordination accuracy in the digital assembly environment with a systematic control method, on the basis of each error item in the assembly process, the following detailed engineering problems are found or need to be resolved based on site investigation and literature review.

① Which regions or features of the aircraft products need coordination accuracy control?

② How to keep the coordination features in a stable statistical control state as the assembly process progresses?

③ How to achieve the complete control of the assembly object for consistency?

④ There is a large number of mating surfaces for curved parts in aircraft. In the existing literature, the calculation of coordination accuracy between curved surfaces is limited. The complex assembly hierarchical relationships and the evolution of the coordination relationships at different assembly stations are also less taken into consideration.

⑤ How to express the practical status of the coordination regions, the constraint status between different coordination regions, and the transfer process of coordination relationship?

⑥ How does a specific coordination relationship affect the product's final assembly accuracy in the assembly process?

⑦ The accumulation of coordination accuracy is mainly based on a simplified linear dimensional chain calculations, and the research on nonlinear problems is insufficient.

⑧ The deformation error caused by assembly forces is less considered as guaranteeing coordination accuracy.

⑨ The calculation of coordination accuracy is less concerned with the characteristics of multi-station, multi-hierarchy, and multi-reference transformation during the assembly process.

⑩ The analysis of positioning error in flexible assembly tooling and the methods of error guarantee are less studied.

⑪ What is the relationship between each coordination control link and final assembly coordination accuracy in the overall manufacturing process?

⑫ How does the relocation error affect the assembly accuracy?

⑬ What is the influence relationship between the practical established tooling coordinate system and the assembly accuracy of the product?

⑭ In order to achieve the measurement of product assembly accuracy, how should the laser tracker be arranged? What are the principles and optimization methods for the arrangement of the ERS (enhanced reference system) points? How is the fixture coordinate system built from the TB (tooling ball) points? How to optimize the OTP (optical tooling point) for distribution on tooling equipment and products?

⑮ How to map the elements of process coordination, such as machining/tooling/measuring elements, according to the product's coordination factors?

⑯ How to quickly design corresponding tooling equipment based on product design information?

⑰ What is the coupling relationship between different kinds of error sources? What is the transfer relationships between errors?

⑱ How to analyze the corresponding coordination methods and coordination accuracy for various typical coordination relationships?

⑲ How to balance the relationship between manufacturing accuracy and assembly coordination?

1.1.2 Technical Analysis and Research Requirements of Aeronautical Thin-walled Structure Assembly

With the visits, investigations and practical engineering practices in aviation manufacturing enterprises, it has been summarized that there are significant dynamic change characteristics due to the assembly of weak, rigid, and thin-walled structures. The above characteristics can be described as follows.

Firstly, the number of parts does not increase steadily as the assembly process progresses, and the structural stiffness of the intermediate assemblies also increases unevenly. Moreover, during assembling the structures, due to the force and shape closure, and the constraint release phenomenon before the final component is removed from the fixture or jigs, the stiffness matrix may has mutations. The above factors cause the influence principle of stiffness matrix changes on assembly performance difficult to obtain.

Secondly, for different assembly procedures and stages, the multiple physical field factors, such as the parts' manufacturing geometric error, assembly loads, and tooling's displacement deviations, have a complex interaction relationship with the assembly local deformation, the internal stress/strain field, and the overall assembly deformation. This coupling relationship would make the assembly deviation accumulation and the internal stress distribution have a dynamic evolution characteristics as the assembly process progress, making the assembly performance very difficult to analyze for thin-walled structures. Additionally, the geometric accuracy and the corresponding internal stress demonstrate a contradictory relationship in the joined structure. Under the dynamic influence of various practical assembly factors, this phenomenon also makes the assembly performance difficult to balance and control.

Thirdly, for the time-varying state parameters on practical assembly site, factors such as the assembly environment, the weight of intermediate assemblies, the locating and clamping positions of the end locators, and cipping, drilling and joining load, make the assembly performance act in a random changing status, resulting in a high level of uncertainty of assembly quality.

During the assembly process, due to the lack of understanding of the dynamic coupling principle for the structure's geometry/mechanical state, the following undesired

Digital Assembly Coordination and Quality Controlling Technology
for Aeronautical Thin-walled Structures

situations often appear on the practical assembly site. This include assembly interference or out-of-tolerance phenomenon, numerical value and area of coordination and repair cannot be predicted in advance, the difficulties of controlling internal assembly stress, and so on. Moreover, due to the factors of assembly deformation, rebound error[5], and the accuracy differences in assembled structures that removed from the fixture, secondary positioning difficulties could occur. Then the remedial compensation solutions, such as repair on the gap and local forced assembly, should be taken for the accumulated assembly deviations. For example, once assembly problems occur between the cabin door components and cabin's main structure, such as profile flush/step, gap, and coaxial error, the workers need to first remove the cabin door from the jig. With the manual grinding and repair work according to their experience, the door would be relocated on the assembly tooling to measure its shape until a qualified profile is gained. Then the door also need to be relocated onto another final assembly jig, for the purpose of continuous coordination for the mating between the suspension joint and profile shape of the cabin's main structure. Only when the "profile" meets the design requirement, can the repair work stop. It is known that the above cumbersome process delays the assembly cycle beyond the expected working period seriously, and the internal assembly stress also cannot be estimated accurately, increasing the risk of flight service. Therefore, predicting and controlling the assembly quality accurately during multi-assembly stages considering the dynamic change characteristics such as assembly structures, geometric accuracy, internal stress, practical assembly parameters, and other assembly factors, is key to reducing assembly uncertainty.

The guarantee of assembly performance parameters, specifically geometric accuracy and internal interactive stress, has become the bottleneck problems that restricting the development of the new-generation aviation products. These products have high-perfor-mance characteristics, such as high stealth, long flying range, long service life, and so on. Correspondingly, the service performance parameters, such as flying safety, reliability, and mobility, have higher design requirements[6]. In their assembly work, key quality indexes, such as profile flush, gap, and the convex-concave value of the fastener joint's head, have been enhanced nearly an order of magnitude compared to previous aviation types, namely, the assembly quality requirements have reached the sub-millimeter level. However, the assembly coordination accuracy is generally tighter than individual parts' manufacturing error[7-9]. For example, ① the assembly step between different skin panels along the flight direction is permitted in modern aircraft, and ② the gap between different skin sections is smaller than the forming/manufacturing accuracy of the sheet parts, especially for CFRP (Carbon Fiber Reinforced Polymer) parts with relatively low manufacturing accuracy. As a result, the guarantee on assembly and coordination accuracy is highly concerned by the design and manufacturing departments.

With the development of 3D design and manufacturing technology, the assembly equipment is undergoing an upgrading period of meeting "functional demand", "performance demand", and "intelligent demand". High-precision drilling robot[10], flexible automatic docking and fine machining equipment[11], and digital measurement devices[12] have demonstrated applications in the engineering production. The aviation manufacturing model also changes significantly, including ① transformation from analog standard gauge to digital coordination model, ② transformation from single measurement to composite measurement representation, and ③ transformation from quality inspection to assembly process detection. However, even with the support of advanced hardware equipment under this "new manufacturing normal" situation, it is still difficult to meet the requirements of cross-generation and upgrading improvement for assembly quality. For example, the profile step between the irregular shape of two titanium alloy cylinder sections can only be guaranteed with passive manual repairs by fitters, with a scrap rate of the whole component caused by the incoordination situation up to 50%. This phenomenon can be called as "cast two to guarantee one". For the assembly period node, it is guaranteed even by "three shifts" within 24 hours. Another example is the assembly gap controlling with new compensation method[13], as shown in Figure 1-4. In the forward fuselage assembly of F-35, the traditional manual shimming method results in excessive time and material waste, making it impossible to achieve a peak production rate of one aircraft per day. As a result, a specially designed tooling prototype system for liquid shim compensation was designed to allow the excess liquid to be squeezed out. In summary, it is urgent to develop the new assembly theories and methods that suiting for the new aviation products and new assembly process equipment, and to explore the assembly performance and quality control mechanism of the aviation structures according to their new requirements.

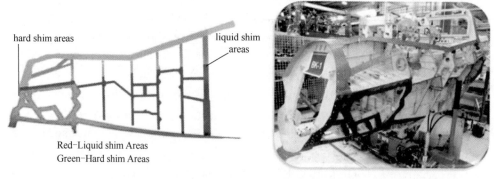

Figure 1-4　F-35 forward fuselage shimmed surfaces[13] (See the color illustration)

With the development of information and communication technology of the next generation, the mode for fundamental reform of manufacturing production is accelerating, and the assembly technology for aviation products is also rapidly moving towards the digital and intelligent era. The development of digital modeling and simulation, data learning and analysis, and experimental testing technology provide new ideas and research tools for the breakthroughs in assembly bottleneck[14]. Then combining with detailed assembly procedures, the assembly methods of the new generation of aviation products should have typical characteristics of "intelligent/ scientific, and active/collaborative control over geometric shape and physical performance", namely:

① Focusing on physical performance control of the assembly structure, i.e. "the combination of geometric shape and internal interactive stress";

② The design and evaluation of assembly parameters has quantifiable characteristic based on scientific design that considers practical measured data;

③ The high-fidelity virtual assembly prototype is taken as the basis of analysis and modeling, reducing the number of physical assembly tests and simplifying the test environment;

④ With the help of modeling and simulation means, assembly performance, shimming/grinding area, and repair amount for different assembly schemes and parameters, can be verified in advance within virtual assembly environment, to apply the closed-loop feedback correcting strategies to control the assembly process accurately.

1.1.3 Structure of this Chapter

For the ultimate goal of reducing the assembly out-of-tolerance phenomenon and rebound deformation in complex thin-walled assembly structures, the academic development context and existing problems for structure assembly were reviewed and analyzed, including assembly process parameters optimization, assembly error transfer and accumulation, comprehensive adjustment to assembly quality, and virtual assembly simulation validation. And benefit efforts would be exhibited, such as the reduction of the uncertainty of geometric accuracy and stress distribution parameters.

1.2 Assembly Process Parameters Optimization

The assembly of complex and lightweight thin-walled structures mainly includes the following working procedures: positioning and clamping operations on fixtures or jigs, drilling holes, fastening and jointing, measuring position and posture, removing from the fixtures or jigs, and so on. The process parameters at each stage directly affect the assembly structures, more specifically, reflected on the value size and balanced distribution of the geometric accuracy and the internal stress. The above items of assembly performance would affect the mechanical and service properties of the assembled structures. In practical engineering site, however, the design of assembly process parameters mostly depends on empirical method or trial-and-error methods, lacking of the optimization support of quantitative and scientific analysis. As a result, from the perspective of the balanced distribution of assembly shape and stress, the research on the optimization method of process parameters in a single positioning, drilling, and joining assembly procedure can lay the foundation for quantitative prediction and control of assembly performance, which is the main aim of this chapter.

1.2.1 Assembly Positioning Accuracy Analysis and Positioning Layout Optimization

For the positioning of aviation complex thin-walled parts or structures with weak rigidity, in order to enhance the local stiffness around the clamping area, multiple end locating effectors are generally required to participate in the locating operation, often adopting the "N-2-1" positioning scheme. In the layout design process for traditional fixture, the number and positional distribution of end locating effectors are firstly designed to fit the assembly characteristics of the product, with basis for formulating the distribution mainly based on production experience or technician's experience rather than theoretical optimal deformation calculations. As a result, the positioning layout needs further optimization considering the higher requirements of assembly quality[15].

Based on the elastic plate theory, Zhou[16] deduced the deformation expression of the workpiece under the comprehensive action of gravity and external loads. Where the average deformation of the key measurement nodes along the normal direction was taken as the objective function, and a rapid design method of fixture locating layout scheme based on hybrid particle swarm optimization algorithm was proposed, to ensure the thin-walled parts have a strong deformation resistance during the positioning stage. Then to optimize the N-2-1 layout scheme of the fixture, and to reduce the maximum deformation of thin-plate workpieces under the action of clamping and joining forces, Arunraja[17] proposed an integrated method that combines finite element analysis, response surface method and Taguchi method. Meanwhile, with the aim of obtaining the influence principle of contact force on panel's deformation, Lu[18] established the mathematical model of the contact force between aircraft panel surface and the fixture under the N-2-1 locating method (Figure 1-5), benefiting results such as minimum positioning deformation of the entire thin-walled workpiece.

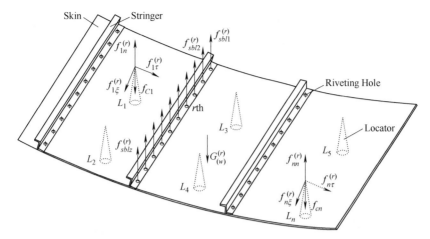

Figure 1-5 N-2-1 locating scheme and contact force analysis as assembling the panel component[18]

$$
\begin{cases}
\text{Find } X = [x_{N1,1}, x_{N1,2}, \cdots, x_{N1,i}, x_{N2,j}, x_{N2,k}, \cdots, x_{N2,n}] \\[2mm]
\min f(X) = \sum_{i=1}^{k} U_i(X) / k \\[2mm]
\text{s.t. } x_i \in \Omega \\[2mm]
\quad\quad x_i \neq x_j \\[2mm]
U_i(X)_{\max} \leqslant \overline{U_i(X)}
\end{cases}
$$

Furthermore, Guo[19] divided the number N into two portions: $N = N_1 + N_2$. N_1 was determined by the actual limitations of the product features; while N_2 was the other positioning points to reduce assembly deformation. As a design variable, N_2 was obtained through the optimization solution under the number and position of constraints associated with N_1. And the searching process for N_2 could be expressed in above Equation. Where Ω stands for the feasible location area of the distributing locating points on parts; x_i and x_j represent different locating points; U_i represents the deformation results of the k key measurement features of the parts; $\overline{U_i(X)}$ represents the range of required assembly deformation; and $f(X)$ represents the objective function related to N_2. Then to determine the accurate locating regions and reduce the deformation, optimization work considering drilling force as well as gravity and locating constraints are carried out on the narrow beam parts.

In summary, for the fixture's positioning layout scheme for sheet metal parts, the above literatures mainly adopt the static analysis method to calculate the maximum local deformation and global deformation. However, only deformation is taken as the optimal objective. This situation would make it difficult to reflect the deformation degree in all directions of key regions, as well as the clamping stress distribution across different areas. What's more, the robustness of the positioning layout, considering the actual manufacturing and clamping states, needs to be further discussed.

1.2.2   Cumulative Effect of Drilling and Joining Deformation and Process Parameters Optimization

In the drilling and joining process, the prediction and control of assembly deformation under different joining parameters is an engineering problem that needs to be solved urgently. These two procedures contain a complex nonlinear phenomena of geometric errors, contact forces and material properties, making it difficult to model and solve. And the relevant researches mainly focus on the analysis of residual stress/strain field caused by joining operations, deformation and load transfer during joining process, equivalent model oriented for joining, and so on.

To find fasteners' configuration with a minimal number of fastening elements that providing the close of admissible initial gaps, Lupuleac[20] carried out the simulation of A320 wing assembly on the base of numerical experiments with the help of ASRP software. Where the assembly processes for A320 were optimized. During the docking of the outer and the central wing box of A350, Zaitseva[21] developed a special simulation program for the riveting process. Where the distribution and number of rivets were designed to get an optimized gap size. By importing the results into finite element software, the joining stress distribution of the assembled object was obtained by post-processing algorithms. In order to predict the deformation of the riveted structure, Zheng[22] established an equivalent mechanical model suitable for the axial expansion process. Where by considering the uneven distribution of radial pressure along the plate thickness, the influence relationship between uneven axial and radial pressure on joining holes' deformation was obtained accurately. By combining with finite element modeling and experimental analysis (Figure 1-6), and based on elastic-plastic and fracture mechanics, Liu[23] obtained the influence of different lapped plate's thickness, rivet sizes, and arrangement distributions on the deformation, residual stress, and fatigue performance of riveted butt joints. In addition, the overall bending effect of the structure caused by the uneven distribution of local residual stress around the interference riveting hole was also analyzed.

Then based on the spring mass model, Liu[24] established the theoretical model of riveted lap joints under tensile loads. With the model, the rivet load stress, diffusion stress, and interference stress around the hole could be calculated. And the influence of rivet's diameter and heading's load on the stress distribution around the hole was then obtained with finite element method. The gained optimized parameters' combination is conducive to improving the fatigue life of the jointed structure. Aiming at the stress/strain characteristics of riveted lap joints, Zeng[25] used the explicit dynamic finite element method to analyze the rivet extrusion process. Where the surface strain changes of the plate were measured with a micro strain gauge, and the distribution state of compression and tension residual stresses in the riveted sheet was also obtained.

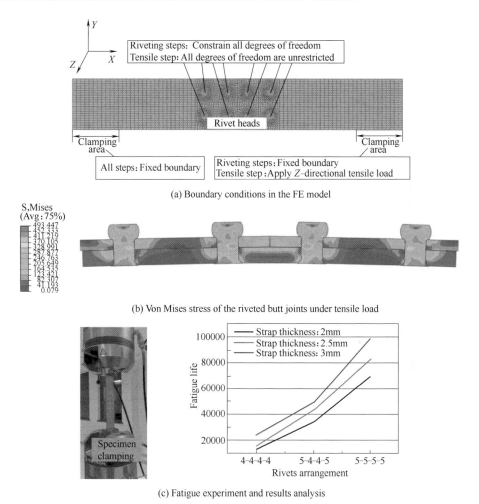

(a) Boundary conditions in the FE model

(b) Von Mises stress of the riveted butt joints under tensile load

(c) Fatigue experiment and results analysis

Figure 1-6　Investigation on the residual stresses and fatigue performance of riveted single strap butt joints[23] (See the color illustration)

From the above researches, it can be known that the prediction of overall deformation of the jointed parts is an important concern point in the field of assembling thin-walled parts. The deformation modeling and numerical simulation in the above literatures mainly focus on the analysis of stress-strain field and the deformation of the riveted structures that containing relatively few fasteners. Correspondingly, the optimization on drilling and joining process parameters mainly takes the local deformation of joining holes as the optimization goal. And there is a lack of efficient calculation method for the joining deformation of large-size skin panels. In addition, solutions can only be gained by carrying out large-scale process experiments making it difficult to obtain the correlation relationship between process parameters and drilling and joining's mechanical quality. As a result, research on the physical simulation technology of these two important processes under the conditions of small samples, would be especially beneficial for optimizing process parameters and reducing the number of tests for advanced drilling and riveting systems.

1.2.3　Fast Finite Element Simulation Considering Virtual-actual Fusion Model

For the parameter optimization during the single assembly process, such as in Sections 1.2.1 and 1.2.2, in addition to the theoretical analytical modeling, finite element method is also often adopted. And this method has been applied in the fields of analyzing the residual stress, structural deformation, and evaluating the process parameters on locating/clamping/drilling/riveting quality. For the thin-walled assembly structures, different parts, such as skins, stringers, stiffeners, and reinforced frames, are riveted or bolted together to form a complete component. However, the problem is that finite element modeling and simulation require a huge amount time, causing this method cannot guide on-site assembly operations effectively in practical engineering.

Aiming at improving simulation efficiency, Lei[26] established different types of finite element model based on force and displacement control method. With this research, the relationship between model sizes, calculation time, and simulation accuracy were balanced, and better results were achieved in riveting parameters and riveting quality control criteria. Then in order to analyze the thin-walled parts' deformation that caused by assembly positioning constraints and clamping forces, Wang[27] established the shell element equation with non-ideal boundary conditions based on the Kirchhoff-Love hypothesis. Where the analytical solution was obtained by Fourier-Galerkin method, and the displacement corresponding to different types of loads were obtained through the introduction of potential function. From another perspective, as assembling the large fuselage sections, an elastic structure model based on combined beam elements was developed by Cheng[28]. In essence, it is an equivalent idealization of the actual complex structure. With respective loads and boundary conditions, the stiffness matrix of the simplified structure model was obtained by summing the stiffness matrices of the beam elements, as shown in Figure 1-7.

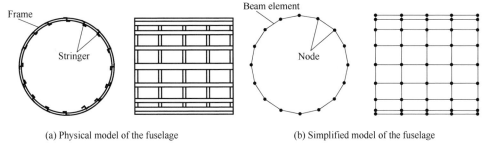

(a) Physical model of the fuselage   (b) Simplified model of the fuselage

Figure 1-7　Elastic structure model based on combined beam elements for fuselage structures[28]

To overcome the limitations of classical structural models and reduce the computational costs, Zappino[29] proposed an approach that could be used to mix one-dimensional beam elements, two-dimensional shell elements, and three-dimensional solid elements together. The refined models is able to deal with the static analysis of complex thin-walled structures. Where the static response of a reinforced panel and a section of an aircraft fuselage was investigated to show the capabilities. For assessing the harmonic response of coupled mechanical systems that involving one-dimensional periodic structures and coupling with elastic junctions, Silva[30] proposed the expression of receptance matrix by means of static modes and fixed-interface modes. However, the above strategies cannot reflect the inter-part deformations that resulted from discrete connection points and overestimat the connection boundary effect severely. Then based on extended Euler-Bernoulli beam theory, Wang[31] derived the consecutive three-dimensional deformation expressions that explicitly describe the nonlinear behavior of physical interaction occurring in compliant component assemblies. It is mentioned that the transformation between the ideal one-dimensional feature and the three-dimensional entity was also introduced. In order to reduce the number of finite element calculation times when modeling the deviations of flexible parts, for repetitive parts or components, with the reduction of the overall stiffness matrix and the substructure's transformation, Lin[32, 33] introduced additional assembly boundary conditions to analyze assembly deviation for the panel subassembly of the side fuselage, as shown in Figure 1-8.

From the above literatures, for repetitive parts or structure features in the modeling process, it is known that with the substructure technology, there is no need to conduct additional finite element analysis, and the prediction efficiency could be improved due to the reduced freedom degrees at discrete nodes. Then for the joining process of body element and shell element contained in thin-walled structures, Chang[34] proposed a free expansion method to eliminate the additional internal stress. Where a three-dimensional dynamic explicit finite element model was constructed, considering the mapping from

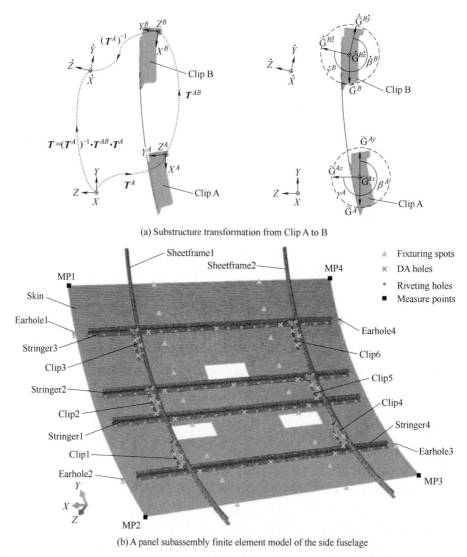

(a) Substructure transformation from Clip A to B

(b) A panel subassembly finite element model of the side fuselage

Figure 1-8 Compliant assembly variation analysis of aeronautical panels with unified substructures[32,33] (See the color illustration)

local bulging to global deformation. And with the help of the equivalent model estimating the plastic zone, the riveting deformation can be predicted effectively. Considering the design and certification process of aerospace structures often requires a detailed stress characterization, for the reduced region of the global model, Fiordilino[35] proposed a global and local modelling strategy to set up an high-order model, as shown in Figure 1-9. Where the first step was to devote the static analysis on a global model of the structure with 1D/2D standard elements, and then a high-order beam model was built locally by

Digital Assembly Coordination and Quality Controlling Technology
for Aeronautical Thin-walled Structures

importing the information from the previous global analysis. As a result, the global structure and local structure have a good continuity in the actual model.

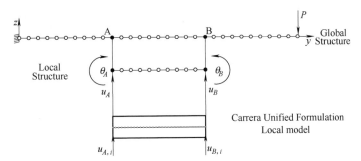

Figure 1-9   Global and local models for the cantilever beam[35]

In summary, after adopting quantitative means to optimize the assembly process parameters, although the above literatures can ensure the positioning accuracy and drilling and joining quality for a single assembly process, the calculation results are quite different from the actual production. This phenomenon could be explained by the ignorance for the practical manufacturing state and assembly constraints, and the simplification on the complex operational links that involved in the single assembly process. Therefore, an effective action, specifically the finite element simulation based on measured model and data, is very essential to address the above shortcomings. Under this situation, the key research contents about the quantitative design of process parameters could be described as follows.

① How to improve the calculation efficiency considering the substructure method?

② How to discretize the mesh element and decompose the region of the practical thin-walled assembly structure?

1.2.4   High Fidelity Modeling for Actual Surfaces of Complex Parts

The realization of the above sections could achieve the displacement compatibility and force/deformation coordination among the interfaces of multi-type finite elements, while the residual stress field could also be reduced. However, for the construction of virtual-real fusion model based on measurement data, it is essential to realize the high-fidelity expression of the actual manufacturing error state for non-ideal parts. It is mentioned that the fusion model could provide the precise model and data basis for the analysis in Section 1.2.1, 1.2.2 and 1.2.3. It is also the basis of accurate analysis of assembly performance indexes in the following chapters.

Considering assembly deformation can be represented by the position errors of the key measurement points, Bi[36] proposed a method based on the D-optimality method and an adaptive simulated annealing genetic algorithm. With these methods, the effective assembly error diagnosis based on the placement of measurement points that cover more deformation information of the panel, could be optimized and realized. Then with the actual data obtained during the dynamic assembly process, Gregorio[37] proposed a hybrid representation method for the product's physical assembly state based on B-rep model, as shown in Figure 1-10. Where the geometry shape of the unassembled components was adjusted by: ① updating the direction and position of the assembled surface; ② moving the mesh nodes of the curved surface. In Figure 1-10, Subfigure (c) is created from the initial as-designed representation of the product (DMU) [Subfigure (b)] and the physical product [Subfigure (a)] so that the extracted manufacturing and assembly operations [Subfigure (d)] result in a final product which fulfills its functional requirements (FR) [Subfigure (e)].

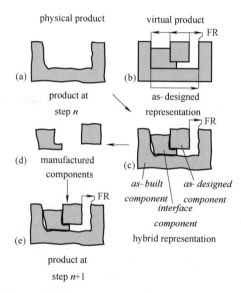

Figure 1-10   The proposed hybrid virtual representation[37]

Combining the availability constraints of the limited actual data and historical measurement data of similar parts, Babu[38] proposed a new Gaussian random field method to express non-ideal part with the complex distribution of surface's normal deviation. In detail, with spatial correlation analysis and condition simulation, the actual error distribution model of 3D free-form surface can be established, as shown in Figure 1-11.

In order to eusure the actual parts consistent with the theoretical model, Hofmann[39] proposed a skin model simulation method based on the principle of product geometric progression specification. Where the geometry shape and dependency relationship of the

Digital Assembly Coordination and Quality Controlling Technology
for Aeronautical Thin-walled Structures

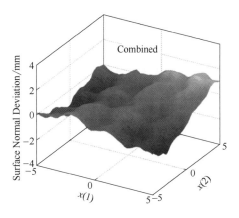

Figure 1-11   Non-ideal part expression with complex distribution model of surface normal deviation[38] (See the color illustration)

workpiece during the intermediate assembly process can be expressed. Then the impact of process steps and manufacturing routes at each assembly stage on the final assembly quality can be quantified. In addition, by extracting the key features of the surface to be measured based on the collected three-dimensional point cloud data, Yacob[40] carried out the anomaly detection work on SMS (skin model shape). With the help of machine learning methods, such as decision tree, k-nearest neighbor, support vector machines, and integrated classifier. Where SMS is defined as a model of the physical interface between the workpiece and its environment, and it is a finite approximation of the infinite skin model. And then the accurate classification on systematic deviation and random deviation is expected to expand the application of SMS in quality control. Considering realistic models regarding with surface geometrical deviations are essential for further functional analysis, Anwer[41] stated that SMS could represent deviations of mechanical parts according to the nature of geometric deviations on the surfaces. And a framework for data-driven geometric deviation generation was also proposed for non-ideal surface modeling in tolerancing, as shown in Figure 1-12. Then from another aspect, by combining the fusion knowledge and traditional assembly variation models, Wang[42] established an effective digital twin-based assembly precision model, with which the accuracy and reliability of the analytical results could be improved.

In summary, by integrating the actual data with the geometric model during the assembly process, the traditional virtual assembly technology can be promoted to approach the assembly for the practical objects. But how to realize the rapid approximation and the high-fidelity modeling of actual deformation surface with complex curvature, and describe the actual distribution of the modified deviation, are the essential problems required further study. Then the obtain of the incremental influence principles

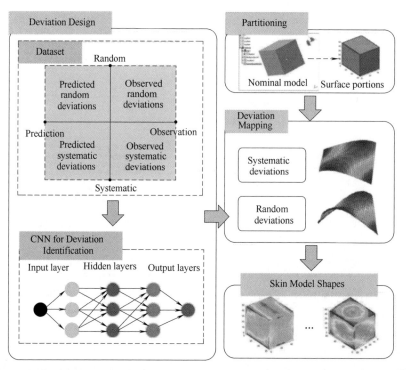

Figure 1-12   Framework of the deviation generation for non-ideal surfaces of SMS[42]

with various assembly process parameters on assembly stress equalization and assembly accuracy control is extremely important for: ① the stability of assemblies' states; ② the quantitative prediction and the active control of assembly performance, especially for the thin-walled composite structures.

1.3   Assembly Error Transfer and Accumulation

Improving the final assembly geometric accuracy, is one of the enduring goals of aviation product manufacturing. Due to the complex mechanics of assembling thin-walled structures, it has difficulties in finding the generation and coupling evolution principles of assembly geometric deformation and assembly stress. The ambiguous geometric shape and internal stress distributions of flexible structures after assembly would result in the final assembly errors that cannot be predicted quantitatively. As a result, considering the high fidelity modeling of actual parts in Section 1.2.4, the research on the transfer and accumulation principles of different error sources can be taken as the basis for realizing accurate prediction and active control of assembly properties.

Digital Assembly Coordination and Quality Controlling Technology
for Aeronautical Thin-walled Structures

1.3.1 Assembly Error Transfer and Accumulation Modeling

For the cumulative transmission modeling of basic assembly error items, considering the geometric tolerances of parts and fixture's positioning deviations, and based on the principles of kinematics, the assembly accuracy modeling for rigid components mostly adopts the method of dimension chain construction, influence coefficient, deterministic positioning, stream of variation, and state-space modeling. However, due to the effects of actual flexible assembly deformation and the dynamic changes of the assembly process, the assembly deviation transfer model based on rigid assumptions has obvious limitations in solving the assembly problems of complex thin-walled structures. In order to enhance the prediction reliability of assembly quality, by combining flexible deformation deviation with actual geometric manufacturing errors, Kang[43] modified and extended the SDT (Small Displacement Tensor) expression of the Jacobian-Tensor model. With the established rigid-flexible coupling model, the tolerance zone and distribution of the target assembly deviation could be obtained, and the assembly constraint relationships of actual flexible body could also be updated with the assembly procedure, as shown in Figure 1-13.

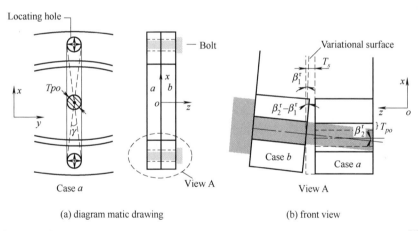

(a) diagram matic drawing   (b) front view

Figure 1-13   Interaction between position tolerance of locating holes and the profile[43]

To prevent adjacent parts from penetrating each other for the influence coefficient, Zhang[44] analyzed the contact effects on dynamic responses of composite structures, where the small scale and simplified isotropic rough surfaces were considered. Tlija[45] considered the dimensions and geometric tolerance of the assemblies by updating the mating constraints of the actual flexible parts. Where the finite element simulations were conducted based on the model that included the distributing state of measured errors. Considering the deviation of mating surface distributing randomly within the tolerance zone, Zhang[46] adopted statistical methods to analyze the uncertainty for non-ideal flat surfaces and proposed a calculation model for the assemblies' uncertains orientation deviations based on gradual adjustment of the coordinate system, as shown in Figure 1-14. Furthermore, considering the non-ideal surface contacting state and based on the established non-ideal SMS models, Zhang[47] adopted the conjugate gradient and fast Fourier transform methods to calculate the deformation of actual surface. Where the deep coupling between geometric errors and contacting deformations could be realized. Additionally, during the multi-station assembly process, Qu[48] established the accurate assembly error propagation model and the quantitative relationship that includes the design parameters of the positioning reference system. Specifically, the discrete-time and non-linear state-space model was adopted and programmed using MATLAB 2014, with motion vector of component-reference system being taken as the state variables.

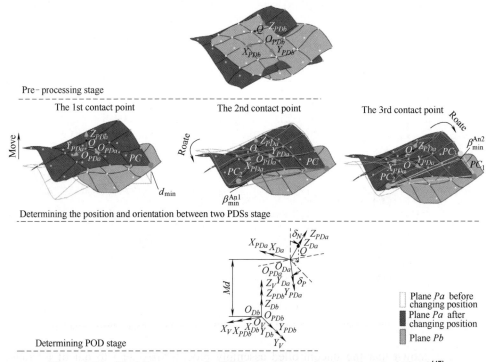

Figure 1-14 Modeling process for PODs (position and orientation deviations)[47]
(See the color illustration)

In conclusion, considering the actual manufacturing information, the above methods provide an accurate representation of the parameter variables that involved in assembly deviation transfer model. However, due to the combined action of multiple process factors, the assembly accuracy of thin-walled structures is significantly affected by the nonlinear cumulative effect among geometric and mechanical deviation parameter sources. As a result, the cumulative effects of various error items with flexible deformation need to be further solved with the view of fundamental research.

1.3.2　Assembly Error Modeling Considering Physical Deformation and Stress

Considering the nonlinear cumulative effects, combined with assembly loads and displacement boundary conditions, Ballu[49] established the relationship between physical assembly characteristics and optimization constraints. Then, considering the difficulties such as ① mating status between non-ideal surfaces; ② model balance under external and internal loads, a simulation method based on linear complementary conditions was also proposed, as shown in Figure 1-15. Finally, the assembly tolerance analysis based on SMS models could be transformed into the objective function of a quadratic optimization problem.

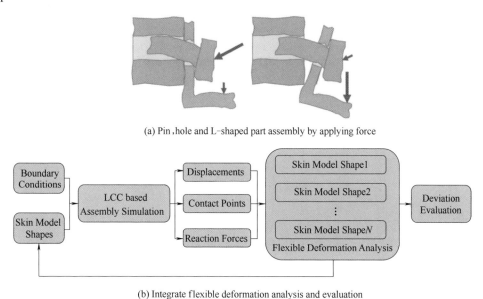

(a) Pin, hole and L-shaped part assembly by applying force

(b) Integrate flexible deformation analysis and evaluation

Figure 1-15　Tolerance analysis using skin model shapes and linear complementarity conditions[49]

It is known that the uncontrolled assembly process may lead to out of tolerance phenomenon, and the accumulation of internal stress will also increase the elastic energy of the entire assembly structure. Aiming at this condition and taking a wing box assembly structure as the research object, Yoshizato[50] proposed a non-linear finite element analysis and optimization method based on element modeling. With the provided accurate and timely assembly adjustment strategies, unexpected assembly deformation and residual stress distribution could be predicted and reduced in advance. Additionally, in order to control the coordinated deformation caused by the internal stress as assembling the large-scale thin-walled structures, Yu[51] constructed an accurate mapping relationship between the basic deformation mode of parts and the spatial deviation field of the assemblies. Also for the wing box component, considering the preload force, Wang[52] constructed the finite element analysis model for the composite thin-walled beam structures with a C-section, as shown in Figure 1-16. With the preload correction, the probability distribution function for the rebound deviation of the reduced scale beam was obtained.

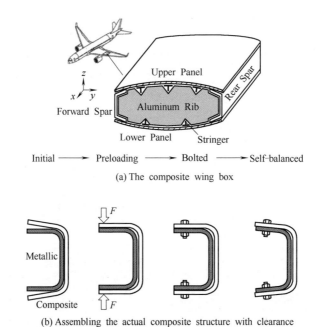

(a) The composite wing box

(b) Assembling the actual composite structure with clearance

Figure 1-16   Wing-box assembly considering preloading-modified distribution[52]

1.3.3   Stiffness Changing in Assembly Error Accumulation

Wrmefjord[53, 54] stressed that assembly deviation management faces challenges such as ① keeping 3D models fully updated, and ② managing the changes and feedback throughout the entire assembly procedures. As assembling the thin-walled structures, considering the cumulative evolution process of assembly internal stress, the structural stiffness related to displacement exhibits a dynamic change phenomenon.

In order to predict the assembly performance that containing physical characteristics and behavioral levels accurately, considering the dynamic characteristics occurring in the assembly process, Stricher[55] improved the influence coefficient method. Where the geometric nonlinearity caused by stiffness changes was taken into account, and the

Digital Assembly Coordination and Quality Controlling Technology
for Aeronautical Thin-walled Structures

influence principles of joining defects, shape tolerances, and rigidity changes on the assembly deviation were gained during assembling the flexible beam component. Then based on the numerical simulations of strain distribution and the calculations of stiffness matrix, Guo[56] adopted the spring model to express the linear equivalence relationship of the actual elastic mating surface, constructing and evaluating the balance criterion that suitable for actual contact conditions, as shown in Figure 1-17.

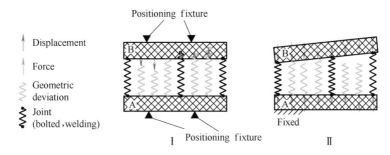

(a) Two types of initial constraint for contact equilibrium status searching

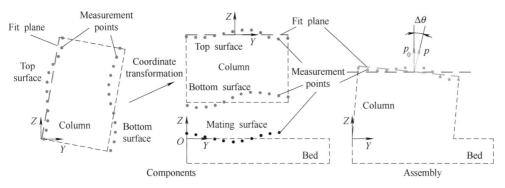

(b) Evaluation of the measured angular error in assembly

Figure 1-17　Assembly variation propagation considering geometric variation and deformation stiffness matrix [56] (See the color illustration)

With the help of nonlinear numerical analysis of contact problems, Kaisarlis[57] discussed the influence of part's stiffness and operation deformation on the functional geometry, proposing an efficient GD&T design method for assembly performance. In addition, for the simulation process of assembling large flexible parts, a lot of calculations relating to different input data should be involved. In order to reflect the meshing simulation results with fine finite element features of assembled structures, Stefanova[58] planned a suitable solver for the quadratic programming problem. Where the equivalent formula for programming was also put forward. However, similar to the influence coefficient method, this method still assumed that the stiffness matrix K and the constraint matrix A were constant values in assembly error models, resulting in the inability to modify assembly simulation model dynamically, as shown in Figure 1-18.

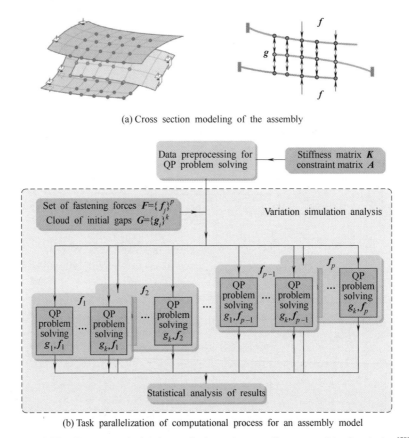

(a) Cross section modeling of the assembly

(b) Task parallelization of computational process for an assembly model

Figure 1-18   Convex optimization techniques in compliant assembly simulation[58]

To sum up this section, the above assembly error transformation and accumulation models generally ignore the dynamic change characteristics during the assembly process, such as the structural stiffness changes, stress stiffening, and internal stress transfer and evolution, as well as their influence on the final error accumulation results. From the specific technical perspective, it lacks the dynamic update mechanism of assembly stiffness matrix and assembly deviation transfer matrix, making it difficult to obtain accurate assembly error and internal stress distribution results. The above research deficiencies would make it not conducive for the process control of assembly performance and fixture's on-site control. Therefore, it is necessary to consider the comprehensive effect of complex multiple physical fields, and the influence of assembly dynamic change characteristics on geometric accuracy, to build a reasonable assembly

Digital Assembly Coordination and Quality Controlling Technology
for Aeronautical Thin-walled Structures

error model. Correspondingly, for complex flexible structures, the out-of-tolerance problem and the rebound stress of assembly performance could be solved.

1.4 Comprehensive Adjustment of Assembly Geometric Accuracy and Internal Stress

Considering the dynamic interaction of multiple process factors, controlling the assembly deformation, springback error, and internal stress of complex thin-walled structures, the dynamic balance and collaborative regulation between assembly geometric accuracy and physical properties is the goal of assembly process optimization. It is also the biggest challenge of achieving high performance assembly.

1.4.1 Controlling Strategy for Balancing Assembly Deformation and Internal Stress

The two kinds of non-independent random variables, i.e. the assembly/coordination deformation and the internal stress, are the results of the collective transmission and evolution of the factors such as structure, load, process parameters and procedures, across both in time and space dimensions. When one performance item is improved, the other performance index will often exhibits a reduction phenomenon. This means there is an inverse growth relationship between these two types of factors. Therefore, to make use of the homologous relationship between shape and force, it is necessary to formulate their cooperative and competitive control strategies firstly. In order to reduce the analysis complexity for assembly stress, based on the methods of unified substructure generating and transforming, deviation propagation modeling, and Monte Carlo simulation, Lin[32, 33] analyzed the internal stress while modeling the assembly accuracy. The internal assembly stress could be recovered from the corresponding assembly deviation using an output transformation matrix, as shown in Figure 1-19.

Then aiming at controlling the gap between parts before joining and to avoid the excessive pre-tightening force for fasteners, the shim's 3D shape that fitting skin panels of vertical tail component was accurately optimized by Franciosa[59]. To be more specific, to satisfy the required assembly accuracy, processing of scanning data and deformable grid model was adopted, and physical characteristics were taken into account to simulate the assembly deviation propagation and the virtual repairment operations among different parts or key regions. In the complex and over-constrained assemblies, although strict tolerances are assigned to parts and their manufacturing/assembly processes, for key assembly features, it is also difficult to meet their different assembly requirements at the same time. Addressing this question, Mckenna[60] correlated the deviation with assembly

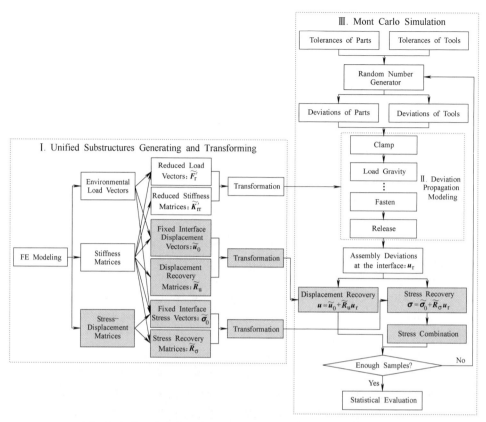

Figure 1-19 Collaborative analysis on geometric deviations and accumulative stresses using unified substructures[32,33]

costs (including repair costs) in different scenarios, allowing the relationship between the manufacturing costs of the entire spar component and the achievable deviation limits to be weighed, as shown in Figure 1-20.

By analyzing the coupling effect and calculating the tolerance distribution parameters between parts of fuselage component using statistical solutions, Iaccarino[61] proposed an innovative assembly method based on the integration of tolerance prediction and DA (Determinant Assembly). Where the number of key features in the assembly and the difficulty of controlling assembly gap can be reduced. Then considering the research on controlling the numerical value and uniformity of assembly stress, Wang[62] evaluated the entropy of assembly stress's location distribution with the strain energy density method. Where the relative entropy was taken as the stress distribution difference index between the ideal model and the actual model that containing error information, while the maximum stress that is lower than the average value of all models, were taken as the constraint condition.

Digital Assembly Coordination and Quality Controlling Technology
for Aeronautical Thin-walled Structures

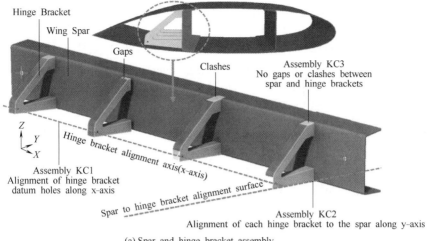

Hinge Bracket

Wing Spar

Gaps

Clashes

Assembly KC3
No gaps or clashes between
spar and hinge brackets

Z
Y
X

Hinge bracket alignment axis(x-axis)

Assembly KC1
Alignment of hinge bracket
datum holes along x-axis

Spar to hinge bracket alignment surface

Assembly KC2
Alignment of each hinge bracket to the spar along y-axis

(a) Spar and hinge bracket assembly

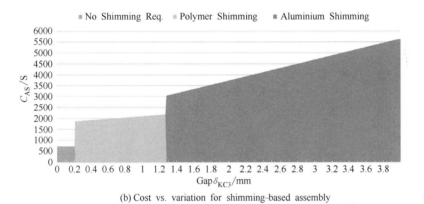

■ No Shimming Req.   ■ Polymer Shimming   ■ Aluminium Shimming

(b) Cost vs. variation for shimming-based assembly

**Figure 1-20   Cost-oriented process optimisation through variation propagation
management for aircraft wing spar assembly[60]**

In conclusion, the above studies on the value size and uniformity of assembly stress
is deepened. However, for complex thin-walled structures, under the mixed constraints of
force and displacement, such as the actual manufacturing error, positioning error, and
assembly loads, the geometric shape and physical properties are constantly coupled
and evolve throughout the assembly process. From the following perspectives of
① numerical level, ② uniformity of distribution, and ③ dynamic evolution, building
a collaborative optimization model of physical properties based on measured data and the
virtual-real fusion model, is the key to realize the dynamic balance in assembly
performance. The balancing strategy should also be described as follows: "Give priority
to ensuring assembly coordination deformation, with reducing the internal stress as the
supplement."

In addition, considering the huge enhancement in assembly requirements for new aviation products, such as the sub-millimeter requirement, and the assembly accuracy requirements exceed the manufacturing errors of individual part. And for assembly coordination accuracy, the requirements might be even tighter. Besides the optimization of positioning and joining parameters during a single assembly process, it is even more necessary to obtain the required target values for assembly accuracy and assembly stress accurately. To achieve this goal, the specific implementation methods for these above adjustment strategies mainly include two aspects: ① the process tolerance parameter control focuses on assembly performance, integrating deformation and internal stress analysis; ② force and position hybrid control for assembly tooling positioning considering time-varying parameters on practical assembly site. Firstly, driven by assembly error transmission mechanism modeling and based on small sample data, preventive control of process tolerance can be achieved through identification on key error links and dynamic and accurate tolerance allocating. Secondly, by integrating the positioning error compensation model with the time-varying working condition data items, the adaptive control of force and position is carried out of the end locating effectors to achieve dynamic compensation and rapid adjustment in positioning accuracy. Then the assembly force between products and tooling, as well as the internal stress among jointed parts could be reduced. With the above preventive and adaptive control solutions, the guidance for manufacturing and assembly operations, as well as the accurate and minimum amount of shimming and grinding work could be gained. These mentioned contents are to be analyzed in the following four sections.

1.4.2 Preventive Control with Optimizing and Allocating the Assembly Process Tolerance

For the preventive control of process tolerance before assembly, it is mainly considered that with the continuous improvement of assembly performance indexes and the product service reliability, the tolerance allocation for key assembly process links has become one of the most concerned problems for technicians. And for the complex assembly error transmission system, the core of process control is to assign the key assembly error links with the optimal accuracy values to ensure easy manufacturing and guarantee assembly accuracy. As a result, the relevant analytical work before the actual assembly operations has a strong research necessity.

With the analysis of traditional tolerance allocation and optimization model, the calculation basis for dimension chains is mainly based on extremum and probability methods. For these methods, the measured actual error data and the deformation caused by assembly loads are not considered. As a result, the calculated results for each error link

Digital Assembly Coordination and Quality Controlling Technology for Aeronautical Thin-walled Structures

are interval values of error distribution rather than specific optimal solutions with specific values. Under this situation, the accurate prediction on assembly accuracy cannot be realized. The solution result is a discrete distribution, making the quantitative and controllable goal cannot be achieved. Similarly, the historical tolerance schemes or industry standards for finalized products are often taken as basis, leading to repeated repairs in mating/coordination areas and making it impossible to guide the new aviation product's assembly operations in practical engineering.

For the specific tolerance optimization with a detailed numerical value size, firstly, to solve the tolerance allocation problem by quantifying the dimensions' manufacturing difficulties and minimizing the tolerance costs, Ghali[63] proposed a modeling method that integrates failure modes, cumulative effects, and threshold analysis tools, as well as a new tolerance allocation method using Lagrange multipliers. Then, considering the factors of manufacturing cost and quality loss for the cabin door component, Jing[64] established a cost tolerance model with the hybrid optimization by Monte Carlo simulation and adaptive differential evolution, as shown in Figure 1-21. Correspondingly, the global search ability and algorithm convergence of the allocation model were improved.

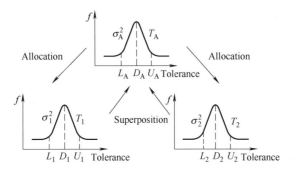

Figure 1-21   The process of tolerance allocation and tolerance superposition[64]

Considering other factors, such as the deformation caused by gravity as a special constraint condition, Fan[65] constructed an optimal tolerance allocation model integrates small deformation items and then solved the model with genetic algorithm. Moreover, considering that the shape error would increase the estimation uncertainty of functional requirements, as balancing the accuracy performance and assembly cost, He[66] proposed a statistical tolerance allocation method. Firstly, by combining the Jacobian matrix with the skin model shape, the assembly state could be estimated accurately. Then, effective strategies of small batch gradient descent and backtracking solution were adopted. The trained deep Q network method was also taken to solve the global tolerance scheme, finally obtaining the optimal tolerance design, as shown in Figure 1-22.

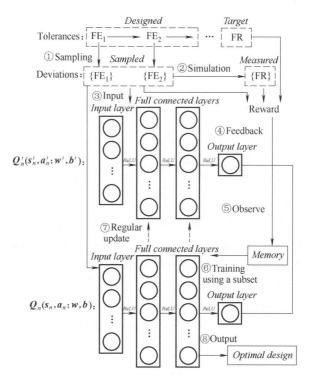

Figure 1-22 Illustration for optimal computation based on deep Q learning for a sub-problem[66]

In summary, the current literatures provide a strong reference role for guiding the modeling and solving tolerance allocation. However, for aeronautical thin-walled structures, the assembly deviation of flexible parts is closely related with: ① assembly coordination deformation, ② internal stress distribution, and ③ accurate error transmission. Moreover, the deformation coordination among different assembly structures would lead to a multi-round distribution of subassembly's deviation across different assembly procedures. Because the consideration for actual physical characteristic data of structural parts and the theoretical prediction results of assembly accuracy model could provide a process data basis for adaptive control in flexible assembly process. Therefore, it is necessary to research the dynamic and accurate allocation laws of tolerance that integrate assembly deformation and actual accuracy/stress distribution status.

1.4.3 Reconfigurable & Flexible Assembly Tooling System and Adaptive Performance Controlling Method

The above contents mainly refer to assembly technology and methodology, however, their realization requires support from the assembly hardware system. From another perspective, considering flexible assembly system refers to tooling with reconfigurable and adjustable function, capable of: ① the automatic force and position collaborative control according to the actual assembly status; ② the assembly of products with different shapes and sizes. As important and widely used production equipment for ensuring assembly performance, its positioning accuracy should firstly be guaranteed. The assembly internal stress is directly reflected in the mating area between assembly tooling and products, caused by geometric and physical factors such as assembly errors and deformation. Considering the combination of assembly hardware system and the corresponding assembly method, to realize the active and adaptive control of assembly performance, the following three typical and representative aspects are to be analyzed in this section: ① the docking assembly and posture adjustment for large components; ② large skin panels with low rigidity; ③ complex structural components.

For the first aspect, based on kinematics design theory, Rainer[67] designed the reconfigurable assembly platform with parallel robots, consisting of multiple kinematic manipulators connected by joints, as shown in Figure 1-23. This handing system had a simplified movement mechanism structure, making it easy to achieve accurate positioning effort within the working range of single operation module, enabling precise and economical posture adjustments for the cylinder fuselage sections and the profile adjustments of the skin panels.

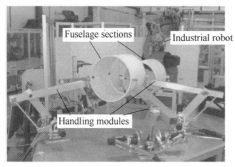
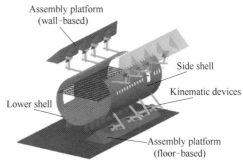

(a) Position and pose adjustment equipment for fuselage sections   (b) Profile adjustment principle

Figure 1-23　Reconfigurable and flexible assembly tooling based on simplified parallel robots[67]

In terms of the collaborative control of force and position, as well as the accuracy assurance for positioning process of assembly tooling, Arista[68] adopted position measurement and force control methods to build an optimal flexible assembly system. Where the physical effects, such as deformation and stress, during the assembly process were considered. The system was applied in the FitFlex assembly project of Airbus A350 XWB, where the strain of rear fuselage side wall panel (14m×5m) was controlled within the allowable limitation. Then, in order to simulate the docking process of the fuselage components (Figure 1-24), Hunt[69] developed a shape control system with multiple reconfigurable end actuators, where the force sensors and deformation measurement devices were contained. And the assembly strategy of releasing the actuator directly rather than applying the reverse force, was also adopted. Then, the finite element simulation of dynamic force curve and virtual assembly analysis were considered. With the designed adjustment process, the dimensional deformation during and after the actual assembly process and the residual stress distributed around the actuators, could be estimated accurately.

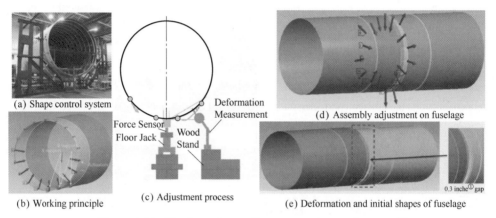

(a) Shape control system
(b) Working principle
(c) Adjustment process
Force Sensor
Floor Jack
Wood Stand
Deformation Measurement
(d) Assembly adjustment on fuselage
(e) Deformation and initial shapes of fuselage
0.3 inche① gap

Figure 1-24   Fuselage shape adjustment and control system[69]

While assembling the recyclable rocket Falcon 9 in Space X, each sub barrel section was positioned on the adjustment bracket, and their relative positions and postures were measured by several laser trackers[70]. With the omnidirectional adjustment function of the supporting frame and the iterative measurement-matching-adjustment process, the closed-loop control of the servo and accurate position adjustments of the fuselage sections could be realized until the axial and radial deviation met the requirements. The application effect showed this flexible automatic docking system could improve assembly efficiency and docking accuracy greatly, while also benefiting the efficient maintenance and repeatable assembly needs. To describe the docking process with more details, Ke[71] designed a posture alignment and joint test system for fuselage components, as shown in Figure 1-25. It can be seen that the fuselage A was placed on the bracket A, and the fuselage B and the bracket B were fixed and joined with four bearing supports. A camera was installed above the docking area of the fuselage, and the real-time display of fuselage docking status could be observed through large screen. Additionally, the laser tracker was

---

❶ 1inch = 25.4mm。

used to measure the four key points set on the fuselage A, and then the integrated management system would calculate the spatial position of the process ball head mounted on the fuselage. When the control system droved four flexible localizers directly below the corresponding process ball head, the fuselage A was lifted synchronously. Then, with the measurement of the actual position and orientation of the fuselage A after the translation and rotation the $X/Y/Z$ direction, the internal adjustment force caused by multi-axis motion coordination error was controlled within the allowable range, ensuring the components' safety in the assembly process.

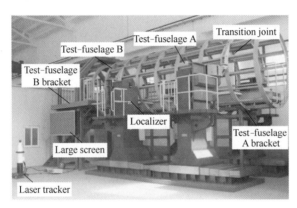

Figure 1-25　Posture alignment and joint test system for fuselage components[25]

For the second situation, involving large skin panels, Mbarek[72] presented holistic solutions for the positioning task, considering operational conditions and environmental influences. The positioning system comprised three portions: measuring system, devices for clamping parts or components, and motion control system. The panel's shape was adjusted by changing the clamping position and force on the panel. Where the movement ability of the tooling system had a total of 48 axes, with three-dimensional force sensors adopted to monitor the assembly forces, aming to avoid excessive internal stress and damage occurs to composite panels, as shown in Figure 1-26. For the large composite skin panel of 20m×6m, the absolute positioning accuracy could reach ±0.1mm.

Then, by analyzing the relationship between part deviation and fixture adjustment across multi-stations, Liu[73] proposed a method for controlling sheet metal assembly dimensions based on assembly fixture's compensation ability. Where the two norm part deviation matrix after fixture adjustment was calculated, and the optimal adjustment values of actuators, considering actual constraints, were gained. With this research, the adaptive control of assembly deviation could be achieved. To be more specific, by dividing the assembly system into two main portions, ① automated joining and flexible holding; ② fastening modular (Figure 1-27), Ramirez[74] proposed an approach suitable for flexible automated assembly systems focused on large CFRP-structures. Where the described automatic processes were machine measurement, contact-point adjustment, 6D-position adjustment with parallel actuators, and shape and force adjustment with vacuum grippers, as shown in Figure 1-28.

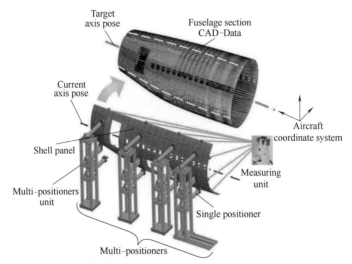

Figure 1-26 Positioning system for the composite shell panels assembly[72]

Figure 1-27 Automated joining and flexible holding fixtures for CFRP panel[74]
(See the color illustration)

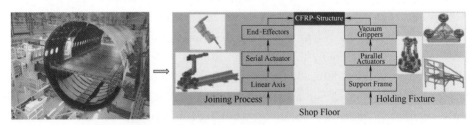

**Figure 1-28 Flexible Automated Assembly System (FAAS) for CFRP-panels
of the Airbus A350XWB[74] (See the color illustration)**

Digital Assembly Coordination and Quality Controlling Technology
for Aeronautical Thin-walled Structures

Unlike the previous two end locating effectors with vacuum cups, Bi[75] realized the adaptive assembly with NC positioner for gaining a better stability and a spacious operational space, as shown in Figure 1-29. Where a special locating mechanism with ball head and socket fitting was designed to locate and adjust the fuselage panel's posture.

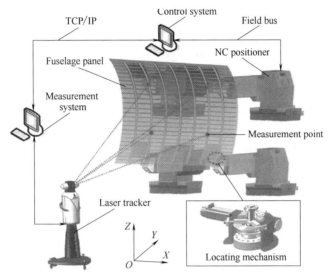

Figure 1-29   Skin panel assembly of the fuselage with ball head and socket fitting[75]

For the third situation, involving complex components, to reduce the extra tooling cost caused by product modification, Jefferson[76] designed a flexible tooling system comprising a mechanical arm and clamping mechanism, and its digital reconfigurable characteristics were verified in the wing-box modification test. To explore the working mechanism of fixture positioning operation on assembly size, Zhang[77] proposed an optimization model and solution algorithm for the fixture's precision compensation along the normal direction. As assembling the wing-box structure, the upper panel was positioned by flexible assembly tooling based on the "N-2-1" principle, with the screws adopted in the clamping mechanism. In addition, its head was equipped with force sensors for monitoring and adjusting the clamping force in real-time (Figure 1-30). Then, the control of assembly deviation based on fixture's active locating compensation, with the compensation effect verified through finite element simulation and physical experiments. In addition, with the invention patents of repeatable processing path using robot and composite material fastening system, the practical problems, such as automatic operation in full size deterministic assembly, were solved[78,79]. The composite vertical tail component of the B-787 was assembled by the flexible system with the shape and force adjustments, as shown in Figure 1-31.

Figure 1-30　Clamping mechanism and physical experiment of wing-box structure[77]

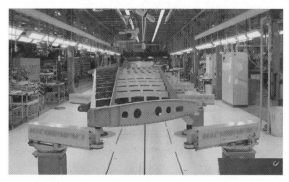

Figure 1-31　Flexible locating system for B-787 vertical tail component[78]

In contrast to the locating method based on contour boards, coordination holes, based on DA technology for locating the 46 ribs of four wing flap components, were adopted by Guo[19]. By dividing the product feature groups, the overall layout relationship among the seven locating units of the flexible tooling system was determined. As assembling these components, the multi-point locating method was taken. Irregular drilling force, overlying gravity, manufacturing error, and internal mating force were all modeled and reflected by force sensors mounted on each locating unit, gaining satisfactory load and position control accuracy.

In addition, the flexible assembly tooling can also be regarded as a special kind of CNC machine tool equipment. For the positioning accuracy guarantee, considering the machine tool's topological structure and low-order body sequence, Yu[80] proposed the concept of relative motion matrix focusing on the connecting support of machine tool's moving axis and the connection matrix of the machine tool's support. Then, the error transfer chain model was established to obtain the machine tool's moving error along the axis. With the above solutions, the machine tool's process adjustment decisions based on the optimal estimation of the error state, could be gained.

In conclusion, compared to CNC machine tools, the positioning of aeronautical

Digital Assembly Coordination and Quality Controlling Technology
for Aeronautical Thin-walled Structures

reconfigurable and flexible tooling has unique characteristics. It is necessary to consider the time-varying state parameters, including tooling's working state, error status of parts and products, environmental state, and assembly force deformation. Through the positioning of different product objects, the dynamic identification on the position/posture adjustment execution parameters for each motion axis of the tooling should be completed. And the intelligent compensation and control methods for solving the accurate positioning problems of flexible assembly tooling should also be developed.

1.5 Virtual Assembly Simulation and Validation Tools and Methods

The virtual assembly technology is developing from the assembly process simulation and modeling, based on ideal geometry models, to physical properties. In the virtual environment, the dynamic interactive analysis of assembly process and assembly internal stress could verify, predict and improve the assembly performance in advance. Benefit results, such as the reliability of the assembly accuracy can be enhanced. As a result, with the introduction of Industry 4.0, the relevant virtual commissioning technologies[80,81] have been paid much attention throughout the whole industry.

To simulate the physical characteristics in a complete virtual environment, Siemens built digital twins focused on product performance in its Simcenter software [82]. Where all of the in-depth analysis results could be fed back to the entire assembly procedures through the MindSphere module, until a completely closed decision cycle was formed. In May 2021, Lockheed Martin[3] applied the "StarDrive" tool, based on digital twins, to complete FSDA (full-size determinant assembly) on composite skin in the development of CHARLIE prototype, as shown in Figure 1-32. The assembly quality can be predicted in advance, and the positioning and coordination processes could be dynamically optimized with the prior virtual simulation. And benefit results, such as the determinant assembly with few assembly fixtures and zero repairs, were gained. It shows the powerful

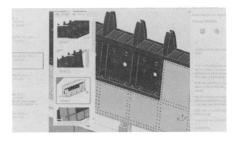

Figure 1-32 The CHARLIE integration verification component in StarDrive plan[3]
(See the color illustration)

ability of controlling the actual assembly process by virtual simulation with virtual assembly prototype.

For the locating and clamping procedure of thin-walled parts, the dynamic clamping force adjustments according to actual deformation was considered by Wang[15], proposing a clamping force control approach driven by digital twin technology. Where the total factor information model of clamping system was built to integrate the dynamic changing factors, along with the virtual space model constructed based on FEM and deep neural network algorithm. The corresponding bidirectional mapping of physical-virtual space for clamping force control is shown in Figure 1-33.

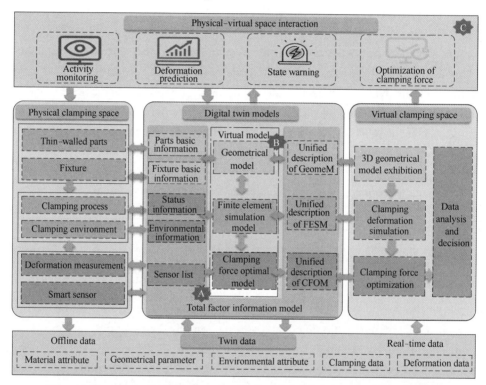

Figure 1-33　Framework for digital twin-driven control of the clamping force[15]

To establish the variation simulation and diagnosis model for compliant assembly, considering deformation, Lee[83] answered how geometric variations were accumulated throughout multiple assembly processes with state space equation, normal equation, and designated component analysis. Where the models were implemented and simulated using CAD tools such as ABAQUS and MATLAB. Then, for the flexible assembly structures, Falgarone[84] developed an assembly deviation analysis platform, named as AnatoleFlex and shown in Figure 1-34. Where the actual physical characteristics were

Digital Assembly Coordination and Quality Controlling Technology for Aeronautical Thin-walled Structures

integrated and taken as the input variables. And the simulation factors, such as assembly sequence, joining mode, material properties, shape deviation, contact modeling, gravity and forces, could be considered to guide assembly tolerance allocation and adaptive positioning adjustment. Finally, the flexible contribution coefficient oriented for the specific requirements could be gained.

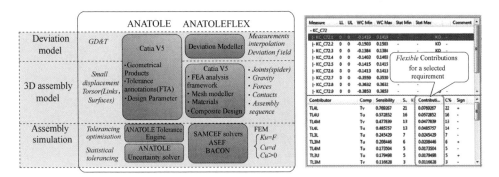

Figure 1-34　AnatoleFlex architecture principle and result analysis[84]

Considering the five-dimensional model within digital-twin technical system that proposed by Tao[14], attentions should be given to the data unity of virtual performance model and the data items that distributed on different assembly stages. Then, the feedback and improvement actions on the twin model of production performance could be done by the accurate detection/measurement on entities. Considering the generation of digital twins of production systems is very expensive, if they are to represent all relevant features of a production system. Beisheim[85] presented that with an IT system of networked software programs and AutomationML container as a special data interface, the digital twins can be generated automatically. To be more specific, the software ISG Virtuos for physical and accurate virtual commissioning was used, and a custom C++ UDP-Interface in Unity3D visualization environment was developed. The use of the Virtuos software integrated into the IT system enables the calculation of the physical relationships among the simulation objects, as shown in Figure 1-35.

As predicting the assembly behavior of thin-walled structures with weak rigidity, Lindau[86] pointed out the challenges encounted during the transformation from physical verification to virtual verification, specifically geometric modeling and computational efficiency. Then, from the perspective of software, combining MSC MAC solver with the MATLAB environment, Corrado[87] developed a tolerance analysis tool called CaUTA (Cassino unified tolerance analysis). Where the finite element software and skin model construction model were contained, as shown in Figure 1-36. By simulating the actual assembly process through the tolerance simulations on rigid and flexible component and analyzing the influence of part's shape errors and mechanical behavior on the super-position results of skin models, the visualization of displacement, deformation, and stress could be realized. The functional requirements could also be verified with pre-processing, online-processing, and post-processing solutions.

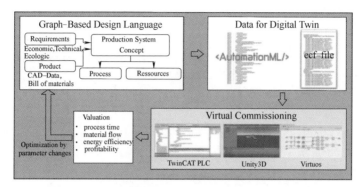

Figure 1-35　IT system for the optimisation of Product, Production Process[85]

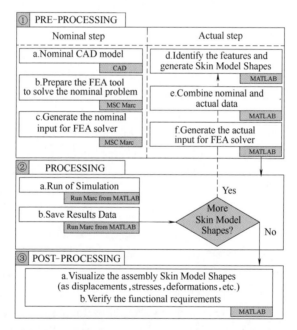

Figure 1-36　Process flow to execute tolerance analysis with CaUTA[87]

In summary, for large aviation thin-walled parts with weak rigidity, their coupling effect of physical properties is extremely strong. Building the virtual assembly platform with the measured model that having a high fidelity and reflecting: ① the stress change state during the actual assembly process; ② the structure's "shape" and "state" intuitively

Digital Assembly Coordination and Quality Controlling Technology
for Aeronautical Thin-walled Structures

and accurately after assembly, are the difficulties for the current virtual verification on assembly performance. In addition, the assembly process parameters and the positioning state of tooling equipment on practical assembly site should also be adjustable in real-time.

1.6　Structure of the Book

As key technology in aircraft manufacturing industry, assembly coordination techno-logy is important for improving the competitiveness and manufacturing level. Considering the above analysis, and aiming to realize the goals of aeronautical digital coordination and quality control for thin-walled structure assembly, the entire book can be divided into six chapters. The main contents and the mutual technical relationships among each chapter is shown in Figure 1-37.

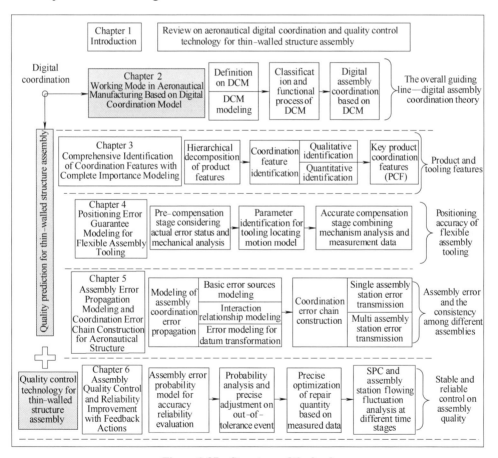

Figure 1-37　Structure of the book

In Chapter 1, with the aim of reduce the out-of-tolerance phenomenon and reduce the uncertainty of assembly performance parameters, the academic development context and existing problems in assembling thin-walled structures are reviewed and analyzed. Key topics include assembly process parameters optimization, assembly error transfer and accumulation, comprehensive adjustment to assembly quality, and virtual assembly simulation validation.

In Chapter 2, to get accurate dimensional size/shape/spatial position and coordination accuracy, according to characteristics of the working mode based on hard master tooling, working mode based on DCM (digital coordination model) is developed. It can replace hard master tooling as the sole basis for the transferring process of dimension/shape/position in aircraft manufacturing. In contrast to the working mode based on hard master tooling, in the measuring field and virtual simulation/analysis environment, the function process of this mode is a dynamic closed loop where comprised CEs/CRs can be continually adjusted in real-time with CMCA. This chapter provides an overall guideline for digital assembly coordination theory.

In Chapter 3, to identity the key characteristic that having a close relationship with the assembly precision, a comprehensive method was developed based on importance calculation. The multi-hierarchy and multi-station assembly process of aircraft products are also taken into consideration, which could bring a higher assembly quality and an enhancement on aircraft's flight performance. This chapter provides the key product features that directly mating with the assembly tooling.

In Chapter 4, considering that the positioning error of the end locating effectors determines the final assembly precision directly, to realize its accurate positioning, a precise and effective error compensation method based on two-stage strategy is proposed. This chapter ensures the accurate positioning accuracy of assembly tooling in practical manufacturing environment.

In Chapter 5, to achieve the goal of precise assembly for an aircraft, by revealing the nonlinear transfer mechanism of assembly error, a set of analytical methods responding to the assembly error propagation process are developed. The ultimate objective is to solve the error problems by modeling and constructing the coordination dimension chain to control the consistency of accumulated assembly errors for different assemblies. This chapter provides an accurate assembly error modeling method and its results.

In Chapter 6, considering the accumulation of different error sources exhibits a complex nonlinear relationship, which causes the frequent occurrence of out-of-tolerance phenomenon and weakens retention ability of assembly accuracy. Novel methods for improving accuracy reliability through probability analysis and prior precise repair compensation before practical assembly operations are provided. This chapter enhances

Digital Assembly Coordination and Quality Controlling Technology
       for Aeronautical Thin-walled Structures

accuracy reliability and ensure assembly accuracy within the required range, laying a precision foundation for product's service performance.

References

[1] Aderiani A, Wrmefjord K, Sderberg R, et al. Evaluating different strategies to achieve the highest geometric quality in self-adjusting smart assembly lines[J]. Robotics and Computer-Integrated Manufacturing, 2021, 71: 102164.

[2] Guo F, Zou F, Liu J, et al. Working mode in aircraft manufacturing based on digital coordination model[J]. International Journal of Advanced Manufacturing Technology, 2018, 76(5-8): 1-25.

[3] Lockheed Matin. How to Build Aircraft Articles in Half the Time[EB/OL]. (2021-5-12) [2024-12-03]. https://www.lockheedmartin.com/en-us/news/features/2021/How-to-Build-Aircraft-Articles-in-Half-the-Time.html

[4] Sun X, Bao J, Li J, et al. A digital twin-driven approach for the assembly-commissioning of high precision products[J]. Robotics and Computer-Integrated Manufacturing, 2020, 61: 101839.

[5] Wang K, Liu D, Liu Z, et al. An assembly precision analysis method based on a general part digital twin model[J]. Robotics and Computer-Integrated Manufacturing, 2021, 68: 102089.

[6] Deng Z, Huang X, Li S, et al. On-line calibration and uncertainties evaluation of spherical joint positions on large aircraft component for zero-clearance posture alignment[J]. Robotics and Computer Integrated Manufacturing, 2019, 56: 38-54.

[7] Bullen G. Automated/Mechanized Drilling and Countersinking of Airframes[M]. Warrendale: SAE International Press, 2013: 1-90.

[8] Williams G, Chalupa E, Billieu R, et al. Gaugeless tooling[C].Aerospace Manufacturing Technology Conference & Exposition, Long Beach, 1998.

[9] Rebello A, Ostrowski M, Yokoyama K, et al. Method and system for creating a tooling master model for manufacturing parts[P]. US Patent, US6856842, 2005.

[10] Mei B, Liang Z, Zhu W, et al. Positioning variation synthesis for an automated drilling system in wing assembly[J]. Robotics and Computer-Integrated Manufacturing, 2021, 67: 102044.

[11] Mei B, Zhu W. Accurate positioning of a drilling and riveting cell for aircraft assembly[J]. Robotics and Computer-Integrated Manufacturing, 2021, 69: 102112.

[12] Li C, Zheng P, Li S, et al. AR-assisted digital twin-enabled robot collaborative manufacturing system with human-in-the-loop[J]. Robotics and Computer-Integrated Manufacturing, 2022, 76: 102321.

[13] Smith J. Concept development of an automated shim cell for F-35 forward fuselage outer mold line control[D]. Wisconsin: University of Wisconsin-Stout, 2011.

[14] Tao F, Qi Q. Make more digital twins[J]. Nature, 2019, 573(7775): 490-491.

[15] Wang G, Cao Y, Zhang Y. Digital twin-driven clamping force control for thin-walled parts[J]. Advanced Engineering Informatics, 2022, 51: 101468.

[16] Zhou S, Qiu C, Liu Z, et al.A Rapid Design Method of Anti-deformation Fixture Layout for Thin-Walled Structures[C]. International Conference on Mechanical Design, Singapore, 2018.

[17] Arunraja K, Selvakumar S, Praveen P. Optimisation of welding fixture layout for sheet metal components using DOE[J]. International Journal of Productivity and Quality Management, 2019, 28(4): 522-558.

[18] Lu C, Huo D, Wang Z. Assembly variation analysis of the aircraft panel in multi-stage assembly process with N-2-1 locating scheme[J].Proceedings of the Institution of Mechanical Engineers, Part C. Journal of mechanical engineering science, 2019, 233 (19-20): 6574-6773.

[19] Guo F, Wang Z, Liu J, et al. Locating method and motion stroke design of flexible assembly tooling for multiple aircraft components[J]. International Journal of Advanced Manufacturing Technology, 2020, 107(1-2): 549-571.

[20] Lupuleac S, Zaitseva N, Petukhova M, et al. Combination of experimental and computational approaches to A320 wing assembly[C]. Fort Worth: SAE AeroTech Congress and Exhibition, AEROTECH 2017, 2017.

[21] Zaitseva N, Lupuleac S, Petukhova M, et al. High Performance Computing for Aircraft Assembly Optimization[C]. 2018 Global Smart Industry Conference, Chelyabinsk, 2018.

[22] Zheng B, Yu H, Lai X. Assembly deformation prediction of riveted panels by using equivalent mechanical model of riveting process[J]. The International Journal of Advanced Manufacturing Technology, 2017, 92(5-8): 1955-1966.

[23] Liu J, Zhao A, Ke Z, et al. Investigation on the Residual Stresses and Fatigue Performance of Riveted Single Strap Butt Joints[J]. Materials, 2020, 13(15): 1-19.

[24] Liu J, Zhao A, Ke Z, et al. Influence of Rivet Diameter and Pitch on the Fatigue Performance of Riveted Lap Joints Based on Stress Distribution Analysis[J]. Materials, 2020, 13(16): 3625.

[25] Zeng C, Tian W, Liu X, et al. Experimental and numerical studies of stress/strain characteristics in riveted aircraft lap joints[J]. Journal of Mechanical Science and Technology, 2019, 33(1): 3245-3255.

[26] Lei C, Bi Y, Li J, et al. Experiment and numerical simulations of a slug rivet installation process based on different modeling methods[J]. International Journal of Advanced Manufacturing Technology, 2018, 97(1-4): 1481-1496.

[27] Wang Q, Hou R, Li J, et al. Analytical and experimental study on deformation of thin-walled panel with non-ideal boundary conditions[J]. International Journal of Mechanical Sciences, 2018, 149: 298-310.

[28] Cheng L, Wang Q, Li J, et al. Variation modeling for fuselage structures in large aircraft digital assembly[J]. Assembly Automation, 2015, 35(2): 172-182.

[29] Zappino E, Carrera E. Multidimensional model for the stress analysis of reinforced shell structures[J]. AIAA Journal, 2018, 56(4): 1647-1661.

[30] Silva P, Mencik J, Arruda J. Wave finite element-based superelements for forced response analysis of coupled systems via dynamic substructuring[J]. International Journal for Numerical Methods in Engineering, 2016, 107(6): 453-476.

Digital Assembly Coordination and Quality Controlling Technology
for Aeronautical Thin-walled Structures

[31] Wang Q, Hou R, Li J, et al. Positioning variation modeling for aircraft panels assembly based on elastic deformation theory[J]. Proceedings of the Institution of Mechanical Engineers, Part B: Journal of Engineering Manufacture, 2018, 232(14): 2592-2604.

[32] Lin J, Jin S, Zheng C, et al. Compliant assembly variation analysis of aeronautical panels using unified substructures with consideration of identical parts[J]. Computer-Aided Design, 2014, 57: 29-40.

[33] Lin J, Jin S, Zheng C, et al. Variation analysis of accumulative stresses in multistep assembly processes using output transformation matrices[C]. ASME 2017 International Mechanical Engineering Congress and Exposition, Tampa, 2017.

[34] Chang, Z, Wang Z, Xie L, et al. Prediction of riveting deformation for thin-walled structures using local-global finite element approach[J]. International Journal of Advanced Manufacturing Technology, 2018, 97(5-8): 2529-2544.

[35] Fiordilino G, Pagani A, Carrera E, et al. Global-local analysis of composite structures[C]. 21ème Journées Nationales sur les Composites, Bordeaux, 2019.

[36] Bi Y, Yan W, Ke Y. Optimal placement of measurement points on large aircraft fuselage panels in digital assembly[J]. Proceedings of the Institution of Mechanical Engineers, Part B: Journal of Engineering Manufacture, 2017, 231(1): 73-84.

[37] Gregorio J, Lartigue C, Thiebaut F, et al. A digital twin-based approach for the management of geometrical deviations during assembly processes[J]. Journal of Manufacturing Systems, 2021, 58: 108-117.

[38] Babu M, Franciosa P, Ceglarek D. Object shape error modelling and simulation of 3D free-form surfaces during early design stage by morphing Gaussian Random Fields[J]. Computer-Aided Design, 2022, 158: 1-21.

[39] Hofmann R, Groger S, Anwer N. Skin Model Shapes for multi-stage manufacturing in single-part production[C]. 16th CIRP Conference on Computer Aided Tolerancing, Charlotte, 2020.

[40] Yacob F, Semere D, Nordgren E. Anomaly detection in Skin Model Shapes using machine learning classifiers[J]. The International Journal of Advanced Manufacturing Technology, 2019, 105(9): 1-13.

[41] Qie Y, Anwer N. Data-driven deviation generation for non-ideal surfaces of Skin Model Shapes[J]. Procedia CIRP, 2022, 109: 1-6.

[42] Wang K, Liu D, Liu Z, et al. An assembly precision analysis method based on a general part digital twin model[J]. Robotics and Computer-Integrated Manufacturing, 2021, 68: 102089.

[43] Kang H, Li Z. Assembly research of aero-engine casing involving bolted connection based on rigid-compliant coupling assembly deviation modeling[J]. Proceedings of the Institution of Mechanical Engineers, Part C: Journal of Mechanical Engineering Science, 2020, 234(14): 2803-2820.

[44] Zhang Z, Xiao Y, Xie Y, et al. Effects of contact between rough surfaces on the dynamic responses of bolted composite joints: Multiscale modeling and numerical simulation[J]. Composite Structures, 2019, 211: 13-23.

[45] Tlija M, Korbi A, Louhichi B, et al. A novel model for the tolerancing of nonrigid part

assemblies in computer aided design[J]. Journal of Computing and Information Science in Engineering, 2019, 19(4): 1-22.

[46] Zhang J, Qiao L, Huang Z, et al. An approach to analyze the position and orientation between two parts assembled by non-ideal planes[J]. Proceedings of the Institution of Mechanical Engineers, Part B: Journal of Engineering Manufacture, 2020, 235(1-2): 41-53.

[47] Zhang Z, Liu J, Anwer N, et al. Integration of surface deformations into polytope-based tolerance analysis: application to an over-constrained mechanism[C]. 16th CIRP Conference on Computer Aided Tolerancing, Charlotte, 2020.

[48] Qu X, Li X, Ma Q, et al. Variation propagation modeling for locating datum system design in multi-station assembly processes[J]. The International Journal of Advanced Manufacturing Technology, 2016, 86(5-8): 1357-1366.

[49] Ballu A, Yan X. Tolerance analysis using skin model shapes and linear complementarity conditions[J]. Journal of Manufacturing Systems, 2018, 48(A): 140-156.

[50] Yoshizato A. Prediction and minimization of excessive distortions and residual stresses in compliant assembled structures[D]. Victoria: University of Victoria, 2020.

[51] Yu H, Zhao Z, Yang D, et al. A new composite plate/plate element for stiffened plate structures via absolute nodal coordinate formulation[J]. Composite Structures, 2020, 247(2): 112431.

[52] Wang H, Liu J. Tolerance simulation of composite wingbox assembly considering preloading-modified distribution[J]. Assembly Automation, 2016, 36(3): 224-232.

[53] Wrmefjord K, Sderberg R, Schleich B, et al. Digital twin for variation management: A general framework and identification of industrial challenges related to the implement-tation[J]. Applied Sciences, 2020, 10(10): 1-16.

[54] Razvan U. Computer-aided Technologies-Applications in Engineering and Medicine Joining in Nonrigid Variation Simulation[M]. London: InTech Press, 2016.

[55] Stricher A, Champaney L, Thiebaut F, et al. Tolerance analysis of compliant assemblies using FEM simulations and modal description of shape defects[C]. ASME 2012 11th Biennial Conference on Engineering Systems Design and Analysis, Nantes, 2012.

[56] Guo J, Li B, Liu Z, et al. Integration of geometric variation and part deformation into variation propagation of 3-D assemblies[J]. International Journal of Production Research, 2016, 54(19-20): 1-14.

[57] Kaisarlis G, Mavridis A, Vakouftsis C, et al. Computational implementation of part stiffness on tolerance specification based on the functional performance of assemblies[J]. International Journal of Advanced Manufacturing Technology, 2020, 111(10): 397-410.

[58] Stefanova M, Minevich O, Baklanov S, et al. Convex optimization techniques in compliant assembly simulation[J]. Optimization and Engineering, 2020, 21(2): 1665-1690.

[59] Franciosa P, Gallo N, Gerbino S, et al. Physics-based modelling and optimisation of shimming operations in the assembly process of aircraft skin panels[C]. IEEE 7th International Workshop on Metrology for AeroSpace (MetroAeroSpace), Padua, 2020.

[60] Mckenna V, Jin Y, Murphy A, et al. Cost-oriented process optimisation through variation propagation management for aircraft wing spar assembly[J]. Robotics & Computer Integrated Manufacturing, 2019, 57(6): 435-451.

[61] Iaccarino P, Inserra S, Cerreta P, et al. Determinant assembly approach for flat-shaped airframe components[J]. International Journal of Advanced Manufacturing Technology, 2020, 108(6): 2433-2443.

[62] Wang Z, Zhang Z, Chen X, et al. An optimization method of precision assembly process based on the relative entropy evaluation of the stress distribution[J]. Entropy, 2020, 22(2): 137-156.

[63] Ghali M, Tlija M, Aifaoui N. Optimal tolerance allocation based on difficulty matrix using FMECA tool[C]. 28th CIRP Design Conference, Nantes, 2018.

[64] Jing T, Tian X, Liu X, et al. A multiple alternative processes-based cost-tolerance optimal model for aircraft assembly[J]. The International Journal of Advanced Manufacturing Technology, 2020, 107(5-8): 667-677.

[65] Fan J, Tao H, Pan R, et al. Optimal tolerance allocation for five-axis machine tools in consideration of deformation caused by gravity[J]. International Journal of Advanced Manufacturing Technology, 2020, 111(1-2): 1-12.

[66] He C, Zhang S, Qiu L, et al. Statistical tolerance allocation design considering form errors based on rigid assembly simulation and deep Q-network[J]. The International Journal of Advanced Manufacturing Technology, 2020, 111(11-12): 1-17.

[67] Rainer M, Martin E, Matthias V. Reconfigurable handling systems as an enabler for large components in mass customized production[J]. Journal of Intelligent Manufacturing, 2013, 24(5): 977-990.

[68] Arista R, Falgarone H. Flexible best fit assembly of large aircraft components. Airbus A350 XWB case study[C]. IFIP International Conference on Product Lifecycle Management, London, 2017.

[69] Wen Y, Yue X, Hunt J, et al. Virtual assembly and residual stress analysis for the composite fuselage assembly process[J]. Journal of Manufacturing Systems, 2019, 52(7): 55-62.

[70] Sandra E. SpaceX launches fifth mission for NRO's proliferated architecture[EB/OL]. (2024-10-30)[2024-12-03]. https://spacenews.com/spacex-launches-fifth-mission-for-nros-proliferated-architecture/

[71] Liu J, Li H, Bi Y, et al. Influence of the deformation of riveting-side working head on riveting quality[J]. International Journal of Advanced Manufacturing Technology, 2019, 102(9): 4137-4151.

[72] Mbarek T, Meissner A, Biyiklioglu N. Positioning system for the aircraft structural assembly[C]. Toulouse: SAE 2011 AeroTech Congress and Exhibition, AEROTECH 2011, 2011.

[73] Liu Y, Sun R, Jin S. A survey on data-driven process monitoring and diagnostic methods for variation reduction in multi-station assembly systems[J]. Assembly Automation, 2019, 39(4): 727-739.

[74] Ramirez J, Wollnack J. Flexible automated assembly systems for large CFRP-structures[J]. Procedia Technology, 2014, 15: 447-455.

[75] Bi Y, Yan W, Ke Y. Multi load-transmitting device based support layout optimization for large fuselage panels in digital assembly[J]. Proc IMechE Part C: Journal of Mechanical Engineering Science, 2015, 229(10): 792-1804.

[76] Jefferson T, Benardos P, Ratchev S. Reconfigurable assembly system design methodology: A wing assembly case study[J]. SAE International Journal of Materials and Manufacturing, 2015, 9(1): 31-48.

[77] Zhang W, An L, Chen Y, et al. Optimisation for clamping force of aircraft composite structure assembly considering form defects and part deformations[J]. Advances in Mechanical Engineering, 2021, 13(4): 155-164.

[78] Fletcher L, Crothers P. Path repeatable machining for full sized determinant assembly[P]. US10691097B2, 2022.04.18.

[79] Hansen D, Simpson B. Conductively coated fastening systems for full size determinant assembly (FSDA)[P]. US11303047, 2022.04.18.

[80] Ugarte M, Etxeberria L, Unamuno G, et al. Implementation of digital twin-based virtual commissioning in machine tool manufacturing[J]. Procedia Computer Science, 2022, 200: 527-536.

[81] Illmer B, Vielhaber M. Describing cyber-physical systems using production characteristics and methodical integration into virtual commissioning[J]. Procedia CIRP, 2020, 97: 272-277.

[82] Siemens. Digital-twin-performance[EB/OL]. (2024-12-03)[2024-12-03]. https://new.siemens.com/cn/zh/markets/automotive-manufacturing/digital-twin-performance.html

[83] Lee J, Choi W, Kang M, et al. Variation simulation and diagnosis model of compliant block assembly considering welding deformation[J]. Journal of Ship Production and Design, 2019, 35(3): 263-272.

[84] Falgarone H, Thiebaut F, Coloos J, et al. Variation simulation during assembly of non-rigid components realistic assembly simulation with ANATOLEFLEX software[J]. Procedia CIRP, 2016, 43: 202-207.

[85] Beisheim N, Linde M, OTT T, et al. Using AutomationML to generate digital twins of tooling machines for the purpose of developing energy efficient production systems[J]. Advances in Transdisciplinary Engineering, 2021, 16: 141-150.

[86] Lindau B, Rosenqvist M, Lindkvist L, et al. Challenges moving from physical into virtual verification of sheet metal assemblies[C]. Proceedings of the ASME 2015 International Mechanical Engineering Congress & Exposition, Houston, 2015.

[87] Corrado A, Polini W. FEA integration in the tolerance analysis using skin model shapes[C]. 15th CIRP Conference on Computer Aided Tolerancing, Milan, 2018.

Digital Assembly Coordination and Quality Controlling Technology
for Aeronautical Thin-walled Structures

Chapter 2
Working Mode in Aeronautical Manufacturing Based on Digital Coordination Model

2.1 Introduction and Related Work

The determination and justification of an aircraft's costs to customers is substantiated by method of its manufacture. The more difficult and complex the means and methods of production, the higher cost of the produce. Airframes have logic production breaks, where individual components are progressively combined and sub-assembled as small units until they form a finished airplane. To get the accurate dimensional size, shape and spatial position, three kinds of working modes are often used in aircraft manufacturing: hard master tooling, digital master model and their mixed use[1]. Hard master tooling is an entity that contains precision information and relative spatial positions of certain parts/components and their related manufacturing toolings, stored in a special workshop for accuracy maintenance. Its appearance is based on the technology development level and available materials situation in the 1940s. However, digital master model is a virtual 3D model emerged in the early 1990s and is stored on computer. The model and the tooling have the same function that they can guarantee the accurate transfer of dimensional size, shape and spatial position of different workpiece from the designed aircraft to the flying airplane.

For a long time, the working mode based on hard master tooling has been an efficient way for manufacturing many kinds of aircraft. But it needs to make quantities of additional special equipment, such as lofting drawings, measuring templates, prototype workpieces, standard metric gauges and dozens of other tools to aid and facilitate the assembly of an airframe. With a serial working method, the master entities are manufactured mainly based on manual work and would take a long period for production cycle[2]. It is mentioned that hard master tooling functions as testing equipment. As manufacturing the transport aircraft Ilyushin Ⅱ-76 with this mode, more than 50000 dedicated jigs or tooling were manufactured, taking nearly 6 times of development cycle for the entire aircraft and accounting for 25% of the total cost[3]. As the requirement for manufacturing accuracy increases, the most lethal problem of this complex hand-built manufacturing method is it cannot meet the precision qualification and manufacturing efficiency standards nowadays. With the popular and deepening application of information technology, digital definitions of products, NC machining, digital measurement and other advanced manufacturing methods are widely used in aircraft manufacturing in digital environment[4-7]. For example, digital tooling[8], 3D-modeling[9], model-based definition (MBD)[10], determinant assembly (DA)[11], design for assembly and manufacture (DFA/DFM)[12], measurement-assisted assembly (MAA)[13], flexible assembly tooling[14], radio frequency identification (RFID)[15], virtual reality (VR)/augmented reality (AR)/ mix reality (MR)/ cinematic reality (CR)[16], digital twin[17], industrial robot[18], assembly line[19], smart factory[20], and more have been researched in recent years, while manufacturing improvements have been put in practical application in companies such as Boeing, Airbus, Bombardier, and AVIC (Aviation Industry Corporation of China). Correspondingly, the working mode in aircraft manufacturing based on hard master tooling is changed to digital coordination model, achieving good results such as economic benefit and manufacturing quality[21-23]. In the manufacturing process of the Boeing B737-800, a mixed working mode based on hard master tooling and digital master model is used on the 48th section of fuselage component[24]. Where the interchangeable entities are manufactured with NC methods and the accumulated assembly error between different assemblies is only 0.0127mm, with parts, tooling, and machine functioning in a harmony status. In the development of B777, hard master tooling was abandoned at more than 3000 assembly interfaces, shortening the production period by 50%, reducing the rework rate by 75%, controlling the costs to 25%, and decreasing the engineering changes by nearly 90%[25]. The savings from the improved assembly method (which reduced touch labor) have reduced price of the plane. With the promotion and alternation of working mode to fit the new normal of digital manufacturing environment, recent application of automated systems of older airframes, such as the T-38 Talon wing in Northrop Grumman, has introduced the need for engineers to translate older paper drawings into 3D engineering models. The T-38 was designed in the 1950s, before computers were available for airframe designers. At the end of production run, paper drawings were scanned into microfiches and stored. The re-winging of T-38 to extend its service life presented a unique challenge: data had to be retrieved from a medium no longer used, and the 2D numeric information need to be converted into

Digital Assembly Coordination and Quality Controlling Technology for Aeronautical Thin-walled Structures

a computer-generated engineering model for NC programming. Furthermore, the overall application of MBD technology in the manufacturing of A380, B787 and C919 aircraft is a milestone, indicating the application of working mode based on digital master model in digital environment has entered a new stage[26].

With the analysis of present technology situations and their applications in aviation industry, it can be found that the working mode based on digital master model has become the mainstream in aircraft manufacture. But while the technologies that used in practical applications and their obtained beneficial effects are introduced, the detailed research process of the realization for the working mode in actual production is rarely donated in current related literatures. Considering the shortcomings and the advantages of the two working modes, taking the coordination route that contains error transfer process into account, the principal and detailed function processes of the working mode based on digital coordination model is proposed in this chapter, providing a precise assembly method and a manufacturing basis in the transfer process of dimensional size, shape, and spatial position for different parts as the single data source. It is hoped to build the application standard based on digital master model in the whole digital manufacturing process of an aircraft.

The remaining sections of this chapter are organized as follows: Section 2.2 presents the method for controlling coordination accuracy; Section 2.3 shows principles of the working mode based on digital coordination model; Section 2.4, practical engineering cases are presented to indicate the methodology's feasibility; while the final section draws conclusions and concluding remarks.

2.2 Coordination Accuracy Controlling Method

Coordination accuracy is defined as the consistency of accumulated assembly errors between two or more assemblies, considering both cumulative direction and dimension size. In general situations, based on the designed product MBD model, by means of certain CNC machine tools, jigs, fixtures, and measuring equipment of the geometry and dimensions in assemblies are formed. As shown in Figure 2-1, if the errors of two dimensional sizes or shapes of different assemblies are consistent to each other, the transfer process of relevant error links must have a relationship with each other. For aircraft products, the above flow-down process can be presented by a coordination dimension chain. In the transferring process, the symbol m represents the number of common error links (from L_{01} to L_{0m}), while n_1 represents the number of error links (from L_{11} to L_{1n_1}) in assembly A, and n_2 represents the error links (from L_{21} to L_{2n_2}) in assembly B. Given that L_A and L_B are the dimensions/sizes that respectively distributing on two assemblies, and they have two accumulated assembly errors Δ_A and Δ_B. If they are progressively assembled into a large assembly, the two flow-down process would be linked together in coordinated route, leading to coordination error, which is donated as V_{AB}.

In the working mode based on hard master tooling, coordination error is guaranteed by interchangeable items that exist as an entity. Although the master tool having a functions as special measuring equipment, it is not digital measurement equipment. The coordination precision information, such as the accurate dimensional size, shape, or spatial position, is limited to the hard master item, making it almost impossible to get the detailed values. As a result, it is very difficult to control assembly process, assembly accuracy, and positioning precision of jigs or tooling. Under this situation, according to the concept of "assembly coordination", with the help of standard master gauge, the consistency of accumulated assembly errors is controlled at final key locating/jointing interfaces or sorting regions, which is defined as "coordination feature". The manufacture process follows a principal known as interrelation. As a result, coordination accuracy only relies on independent error links (from L_{11} to L_{1n_1}, and from L_{21} to L_{2n_2}) in error transferring process, as shown in Figure 2-1. In this working mode, when $m\neq0$, indicating that no matter how imprecise of the errors of the common links are, assembly errors can have a good consistency with standard master items. The common links nearly play a minimal role in the coordination accuracy.

In another case, within the digital manufacturing environment, with the extensive using of digital measuring equipment and CNC equipment, hard master tooling is not used anymore for economic and manufacturing quality reasons. In this working mode based on digital master model, all kinds of parts and assembly tools are manufactured independently, when $m=0$. Coordination accuracy is the accumulation by all of the error links in manufacturing. With the decrement of error links in coordination route and the enhancement of manufacturing precision, assembly accuracy is to be theoretically increased. But the fact is that there are no common error links in this working mode, for the calculated values of coordination error V_{AB}, it is hard to judge which one is more accurate with the above two working modes.

To guarantee assembly accuracy and realize the goal of precise assembly for an aircraft, the principle of coordination accuracy control method is proposed based on the above analysis, as shown in Figure 2-2.

In the control method, where $\Delta_{\Sigma1}$ and $\Delta_{\Sigma2}$ represent the accumulated assembly errors of assemblies A and B, and V represents their coordination accuracy. Obviously, dimensions of $\Delta_{\Sigma1}$ and $\Delta_{\Sigma2}$ are not the same, and their accumulation directions are

Digital Assembly Coordination and Quality Controlling Technology
for Aeronautical Thin-walled Structures

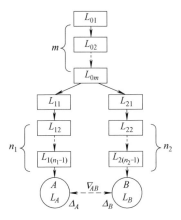

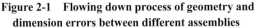

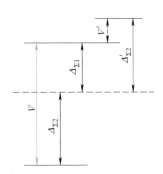

Figure 2-1 Flowing down process of geometry and dimension errors between different assemblies

Figure 2-2 Coordination accuracy controlling method

converse. Then V can be calculated as the sum of two absolute values of $\Delta_{\Sigma 1}$ and $\Delta_{\Sigma 2}$, expressed as $V = |\Delta_{\Sigma 1}| + |\Delta_{\Sigma 2}|$, as shown by the gray bidirectional arrow in Figure 2-2. However, as manufacturing accuracy of parts and forming dies, as well as positioning accuracy of assembly fixtures meet design requirements under digital manufacturing environment, certain digital measuring equipment would be used to monitor the transfer and accumulation of error items during assembly process. If V couldn't meet the requirements, based on the real-time measuring and acquired data from practical assembly site and simulation software, certain coordination relationship would start a feedback adjustment process for controlling the consistency of the assembly errors. As shown by the dashed bidirectional arrow in Figure 2-2, when $\Delta_{\Sigma 2}$ is controlled to $\Delta'_{\Sigma 2}$, which has a converse direction oppsite to each other, coordination error can be expressed as $V' = |\Delta'_{\Sigma 2}| - |\Delta'_{\Sigma 1}|$ with the assumption that the absolute value of $\Delta'_{\Sigma 2}$ is greater than $\Delta_{\Sigma 1}$. By comparing the values of V with V', it can be known that V' is less than V, and this beneficial result is the action of feedback adjustment process on coordination relationship. By controlling the consistency of accumulated assembly errors both on value size and direction, the influence of manufacturing errors on coordination accuracy can be minimized to a certain degree.

2.3 Principle of the Working Mode Based on Digital Coordination Model

According to the manufacturing characteristic of the new normal of digital manufacturing environment and taking the control method for the coordination accuracy into account, principle of the working mode based on digital coordination model (DCM) is proposed, as shown in Figure 2-3. It is mentioned that DCM also stands for digital master model in this chapter, which aims to increase assembly efficiency and optimize coordination accuracy in aircraft manufacturing industry.

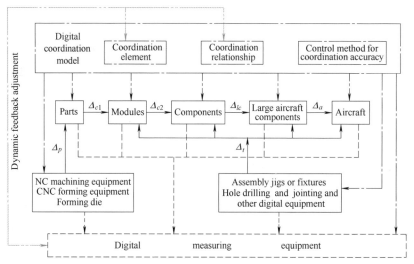

Figure 2-3　Principle of the working mode based on digital coordination model
(See the color illustration)

In Figure 2-3, DCM consists of three main portions: coordination element, coordination relationship, and control method for coordination accuracy. The dotted line arrows (colored in red) represent the measuring operations in the flow-down process for the whole assembly stage with digital measuring devices. The dash-dot line arrows (colored in green) denoting the three action process of DCM on manufacturing activities. The first process is to assemble different parts/modules/components/large components into a whole aircraft. The second is using different kinds of equipment in production, such as NC machining equipment, CNC forming equipment, forming dies for sheet metal parts, assembly jigs or fixtures, hole drilling and jointing equipment, and other digital devices. The third type is to operate of digital measuring equipment on the practical assembly site and within manufacturing sceneries. The above line arrows also indicate the coordination data flowing process for aircraft products and different kinds of equipment originating from one single data source in DCM. Another type of arrow is the double dot marking (colored in violet), indicating the dynamic feedback adjustment process on DCM based on the real-time data from digital measuring equipment. In addition, Δ_{c1}, Δ_{c2}, Δ_{lc}, and Δ_a represent the assembly error items at different assembly stages, while Δ_p stands for the manufacturing errors of parts, and Δ_t represents the locating/jointing precision of the assembly fixtures or other relevant equipment. The above error items have a direct relationship with the control of the consistency of accumulated assembly errors, specifically assembly coordination accuracy.

In the working mode based on DCM, the only basis for the flow of coordination relationship is digital master model. Since none of the interchangeable entities items are put into production, this would lead to a substantially reduction of error links for controlling accumulated assembly deviations. The coordination information exists in the form of digital numerical value that can be easily gained. By quoting the information of coordination elements and coordination relationships from DCM, combining a large number of advanced manufacturing/assembly/measuring equipment, the manufacturing/assembly accuracy of all coordination error links can be controlled to meet design requirements. The specific approach is to analyze and compare the precision requirements that defined in DCM with actual measuring results. When the measured data cannot meet the required accuracy at certain assembly stage, the measured results need to be checked and analyzed firstly, to verify whether the data in assembly process is the same as the coordination data in DCM. Then, with the help of the control method that contained in DCM, precision values of forming die/assembly jigs or fixtures, and design processes of machining/forming/positioning/jointing should be adjusted in real-time or according to the measuring procedure that planned in manufacturing activities. To achieve the purpose of assembly coordination, after calculating the cumulative results of the error links with simulation software, authority of the designed data in DCM should be judged as needed, and precision status of coordination relationship may be adjusted. This dynamic feedback adjustment process is indicated by double dot marked arrows (colored in violet). Then, the method of assembly analysis and simulation can be put into action, and the updated coordination data can be applied to practical production. Lastly, the relative consistency for cumulative direction and value size of accumulated errors between different assemblies could be more easily guaranteed.

2.3.1 Definition of Digital Coordination Model

As the coordination basis in manufacturing process, to guarantee the flow-down process of the dimensional size, shape, and spatial position of different workpieces, there should be an existence of certain precision information and the relative spatial position information as the single date source. For the hard master tooling, coordination basis exists as the text description information and physical entities. In contrast to the tooling, the master items that defined in this chapter exist as a virtual 3D process model, which is saved and stored on computer. It is the only and entire basis for coordination process in production and serves the same function as hard master tooling.

[Definition 1]: digital coordination model (DCM)

DCM is built based on MBD model of products and is expressed as an integrated 3D process model that containing a serial complete coordination information, such as coordinate system, datum reference system, certain key features and parameters of parts/components/forming-die/jigs/fixtures/measuring equipment, and their precision information related with the consistency of accumulated assembly errors. The relative spatial position and other parameters that influence coordination accuracy would also be included. In addition, control methods are also defined in DCM, which play an important role in guaranteeing coordination accuracy between two or more assemblies. DCM can be applied in ① coordination process of parts machining and forming, ② tooling design and installation, ③ assemblies positioning and jointing, and ④ measuring operations for precision/ spatial-position/physical status. The detailed composition can be shown in Figure 2-4.

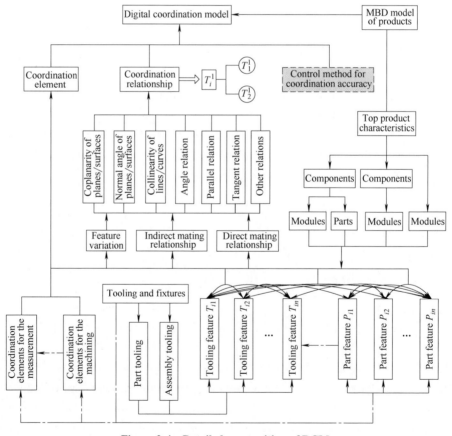

Figure 2-4   Detailed composition of DCM

Digital Assembly Coordination and Quality Controlling Technology
for Aeronautical Thin-walled Structures

It is mentioned that only the detailed composition of DCM is presented in the above figure. However, when the design process for coordination planning and simulating is carried out based on theoretical DCM model, there would be a big difference between the practical assembly coordination state and the theoretical calculation results. This may be attributed to the variations of physical status, such as assembly deformation, positioning error of tooling/jig, and the continuous accumulation combining assembly status changes. To overcome this problem, an efficient solution is to construct the DCM that can reflect practical assembly geometry and physical deformation status. It would be taken as the only basis to control coordination accuracy and could also provide a precise and sufficient digital assembly data to control coordination accuracy. Finally the phenomenon of inconsistency can be avoided or reduced. The data source of DCM can be detailed in Figure 2-5, integrating information of practical mating precision states, position of coordination features, physical deformation during assembly, tooling data, secondary assembly deformation, and more. And it also keeps in accordance with MBD design model and assembly site through digital thread. In essential, DCM is the space mapping of practical assembly site and product assembly entities. It is worthy pointed that DCM is a dynamic model envolves as the assembly process flowing down, with which assembly coordination routine planning work can be carried out more precisely and easily.

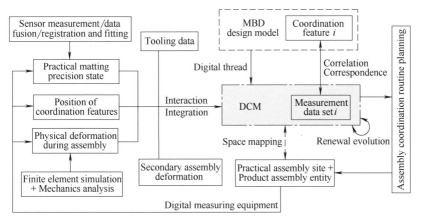

Figure 2-5   Data sources of DCM

As shown in Figure 2-4, the defined DCM is comprised of three portions: coordination element, coordination relationship, and control method for coordination accuracy. Combining with Figure 2-5，these key portions are defined and explained as follows.

[Definition 2]: coordination element (CE)

CE is the carrier of coordination information that contained in dimensional size and shape transfer process of aircraft manufacturing. The change of its geometric size, shape, position or other attributes, such as manufacturing parameters, would affect the coordination accuracy between two or more assemblies. To be more specific, it consists of certain geometric features distributed on parts/components of the aircraft product, and their corresponding assembly tooling/ fixtures. According to the definition of DCM, the machining and measuring parameters in multi stations and hierarchies of manufacturing process could also have an influence on coordination accuracy, and they are correspondingly defined as a member of CE.

[Definition 3]: coordination relationship (CR)

For aircraft products, CR refers to the manufacturing relationship, the inspection relationship, and the assembly/matting relationship among different kinds of parts/ components, and their corresponding manufacturing tooling/fixtures/equipment. In detailed, CR can be taken as the geometric constraint relationship or the precision relationship of dimension/size/position actually, and it is ultimately reflected in the dimensional size, profile shape, and spatial position of assembly bodies. In other words, CR is basis for controlling aircraft assembly and coordination accuracy. With the new normal of digital manufacturing environment nowadays, due to the lack of hard master tooling, assembly accuracy would have a tight relationship with the assembly process of different kinds of parts or components, and the locating and joint interfaces or the assorting regions would change among a series of parts and equipment. The CR in digital manufacturing environment can be illustrated in Figure 2-6. As a complement, based on the variation of CEs, CR can be classified into direct mating relations and the indirect mating relations. The latter is the geometric relationship that existing among parts or components at different assembly hierarchies or stations. However, they both have a more detailed classification into categories such as ① coplanarity and normal angles between planes/surfaces, ② colinearity between lines/curves, and ③ angle/parallel/tangent relationships.

CRs, under the new normal of digital manufacturing environment, present a network structure that is closely related to product structure and assembly hierarchies. In Figure 2-6, the one-way arrow indicates the multistage assembly process of parts, modules, and components of an aircraft with different kinds of equipment, jigs, fixtures, and tooling that provide support function for aircraft components during the manufacturing process. In addition, the above equipment also has a function of locating different components at correct relative positions during assembling. The bidirectional arrows represent CRs among different assemblies or their corresponding equipment, which actually is their mutual geometry/position deviations and constraint relationships. Due to one of the key characteristics of aircraft assembly is multi-level and multi-stage, CR is not only embodied in the direct constraints between assemblies or their related equipment, but also exists among different assembly levels or stations, which means the relationships are indirect. As the manufacturing process flowing down, it is mentioned that these CRs would mix together in a non-linear cumulative process to ensure the assembly/coordination precision of the whole aircraft.

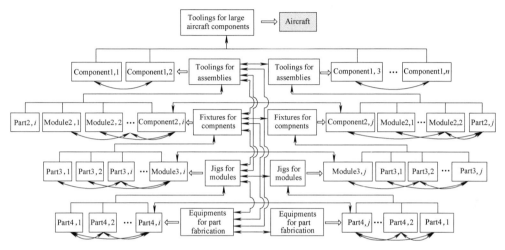

Figure 2-6   Coordination relationship under digital manufacturing environment

[Definition 4]: control method for coordination accuracy (CMCA)

As a comprised portion of DCM, for error items between assemblies, CMCA has a function of ensuring relative consistency of the accumulated value size and transfer direction. However, it is not a visible content in DCM, shown by the dotted frame in Figure 2-4. Instead, it functions to determine the detailed influence of CEs /CRs on coordination accuracy in manufacturing process and to provide technical support for the whole coordination process. The affecting process of CMCA can be summarized in five sections: ① identification and control of coordination elements, ② mapping between product coordination features and process coordination parameters, ③ assembly error propagation modeling and coordination error chain constructing, ④ classification and identification of assembly coordination error sources, and ⑤ tolerance allocation for assembly coordination process. The detailed function process can be indicated by the closed-loop arrows in Figure 2-7, which act as a dynamic feedback adjustment for coordination accuracy.

The first section can be explained as follows: In the working mode based on hard master tooling, CEs are limited in master tooling demonstrably. But in the date flow process based on DCM, coordination accuracy is reflected by the consistency of

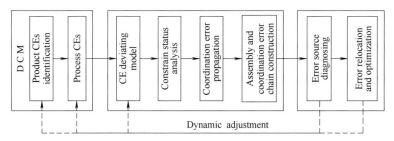

Figure 2-7　Control method for coordination accuracy with DCM

accumulated assembly errors, which has a tight relationship with the detailed assembly process. This would lead to changes on the CEs. To identify the product CEs and enhance the coordination accuracy in digital manufacturing environment, the past engineering experience based on hard master tooling is not suitable anymore. In addition, due to complex structure of aircraft, the whole manufacturing process can be divided into many assembly stations and manufacturing stages. As the manufacturing/assembly processes progress, assemblies in lower hierarchies often act as special parts for higher assembly hierarchies. Taking these complex interactions and evaluation relationships between assemblies into account, CEs for certain components at different manufacturing stages may differ from each other. To sum up, for aircraft products, it is meaningful to identity the CEs to guarantee coordination accuracy in the multi-hierarchy and multi-station assembly process in digital manufacturing environment. To CMCA, one of its functions is to determine which element is the key feature from alternative geometric feature sets distributed on parts/components. To solve this problem, two steps are generally contained in identification process: qualitative analysis based on engineering experience and quantitative analysis based on mathematical modeling, as shown in Figure 2-8, where VPC stands for variation propagation chain. The selected CEs may include certain locating/joint interfaces, mating surfaces, or breakdown interfaces for design and manufacturing. Another function is the error variation control for these features, and the control method is mainly based on statistical process analysis and assembly quality fluctuate analysis at different assembly stations. The above analysis would ensure the identified CEs remain in a statistically controlled status during manufacturing/assembly process.

　　To have a complete coordination accuracy control of CEs in product, other CEs that ① distribute on relevant assembly tooling and fixtures, ② contain machining and measuring process, also need to be determined. The function of the second section is to determine CEs in the manufacturing process, allowing the process coordination information to be quickly mapped out and delivered to practical assembly site from the CEs that defined in DCM. The mapping process can be represented by the dash-dot line arrows in Figure 2-4, which means the functional process from DCM to manufacturing activities. Considering the great importance of the measuring links, the mapping process from machining CEs and tooling/fixtures CEs to the measuring CEs is also included. Taking the detailed assembly operations into account, it is mentioned that the assembly CEs can be decomposed into tooling/fixtures CEs and measuring CEs. The above specific mapping relationships between product coordination features and process coordination parameters can be shown in Figure 2-9. It is mentioned that a whole mapping function model can be constructed to reflect this. In detail, polychromatic set theory can be used to bridge product CEs and tooling/fixture CEs. While, combined with the research of

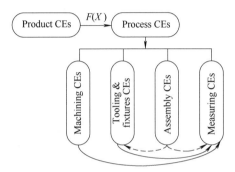

Figure 2-8   Mapping relationship among different CEs

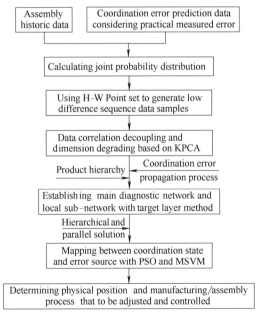

**Figure 2-9   Classification and identification method of assembly
coordination error source under small sample**

selecting measuring equipment, establishing measurement benchmarks, and optimizing measuring sampling/layout/path/sequence, the mapping functions form product design coordination features, machining parameters, and tooling/fixture coordination features related to the measuring CEs can be constructed.

The third section of CMCA is coordination error modeling, which includes the following items: error modeling, error interaction relationship, error modeling in coordination datum transformation, assembly precision propagation at a single station, assembly precision propagation at multiple stations, and coordination chain construction, as shown in Figure 2-10. Analysis in this section is based on detailed CRs in aircraft manufacturing process. As spatial position and posture of assembly structure are determined by the accumulation of geometric data and constraint types of CRs, the precision status of CEs, geometric constraint state among CEs, and flowing down process of CRs should be expressed in DCM firstly with the identified CEs in products and the mapped CEs in manufacturing process. Precision status of CRs consists of fabrication error of parts, installing error of jigs/fixtures, and deformations occurring in assembly process. Values of the first two basic error items can be gained through certain type of measuring operations, and they often obey a normal probability distribution $\xi_p\text{-}N(\mu_p,\sigma_p)$ according to the experience and statistical analysis in manufacturing. For deformation error item, it can be gained according to finite element simulation method based on practical working constraints or conditions. Then, in the propagation process of assembly errors, the above three kinds of basic error would interact with one another. Manufacturing error would combine with positioning error to form the mating error δ_q, which is mainly caused by geometric dimension variations. However, for specific assembly stages at different assembly stages or working procedures, mating error δ_q at previous assembly stage may act as an input error for the subsequent assembly deformation error ξ_d at current stage. Correspondingly, ξ_d at previous assembly stage may also plays the role of an input error for another mating error. To analyze assembly coordination precision accurately, the mutual influence relationships of these basic errors are modeled in CMCA by constructing error influence matrices. As assembly process flows down, different kinds of coordination datums, such as machining datum, locating datum, assembly datum, and measuring datum, often change to fit different working conditions. Correspondingly, actual spatial position of key features in practical assembly work would have additional variation sources contrasting to the theorical position that defined in the product MBD model. In CMCA, error items in coordination datum transforming process are calculated, including modeling practical establishment errors of assembly coordinate system, actual spatial positions of parts or components, and actual relocation error modifications as these CEs are put into practical assemble work.

Additionally, error propagation processes for a single assembly station and for multiple assembly stations are precisely modeled in CMCA to gain coordination error chains for different assemblies. Then the final coordination error can be constructing by accumulating assembly errors across different matting assemblies.

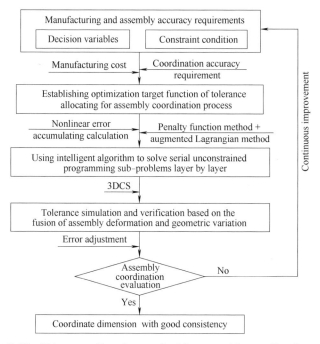

Figure 2-10   Tolerance allocation method for assembly coordination process

To realize the goal of precise assembly, it is mentioned that, although the precision status of CRs and assembly deformation errors are measured and simulated according to practical assembly conditions, the theorical results of coordination accuracy based on CMCA and the actual measurement results should also be analyzed and compared with each other in real-time. If the results are not in accordance with design requirements, then the CEs and CRs would be adjusted within a dynamic feedback process until a good consistency of assembly accuracy is occurred between assemblies, as shown in Figure 2-3. The closed adjustment loop is to be discussed in detail in the next two sections.

The fourth section of CMCA is the classification and identification of assembly coordination error sources, as shown in Figure 2-9. Considering characteristics of multi-level assembly process, attributes of multi-decision variables, and nonlinear transfer processes of coordination errors, the hierarchical diagnosis and identification method of coordination error sources should be studied when coordination requirements are not satisfied. Firstly, according to the assembly history data and the predicted result of coordination accuracy that orienting to practical assembly field, Bayesian network is used to calculate their joint distribution probabilities. Considering the characteristics of "small sample, high dimension, and strong correlation" of the assembly quality data, H-W point set is used to generate low-difference sequence samples. Then, the sample points can be uniformly distributed in the sampling space, avoiding the overlap phenomenon in small fields. Secondly, considering high correlation of various coordination error items, multivariate statistical analysis is carried out. Kernel function with Gaussian orthogonal basis is used in principal component analysis (KPCA), and then the eigenvalues and eigenvectors of data samples are calculated to reveal the correlation between these multi-source variables. Lastly the correlation decoupling and dimensionality degradation of error sampling data can be accomplished by replacing the whole data sample values with fewer independent variables. Thirdly, based on the interaction mechanism among different error sources and the multi-level accumulating principle of coordination errors, target layer method is used to construct the main diagnostic network and the local individual sub-network. Then, the problem of diagnosing assembly coordination errors would be decomposed into a series of parallel solvable sub-problems. Based on particle swarm optimization (PSO) algorithm, the identification parameters of support vector machine (SVM) can be optimized, and then with the intelligent classification and prediction learning algorithm according to Multivariable SVM, the hierarchical mapping relationship between coordination state and error sources could be explored. At last, the combined classification of various error sources that generate assembly quality problems can be solved, specifically the establishment of diagnosing pattern from higher level of assembly station modes to lower level of detailed working procedure steps. Based on the identified physical positions of assembly stations and manufacturing processes which need to be adjusted and controlled, tolerance allocation work for assembly coordination process can be done in the next section.

The fifth section of CMCA is tolerance allocation method for assembly coordination process, as shown in Figure 2-10. By combining multi-level organization of aircraft products with detailed assembly process, considering requirements of "assembly accuracy" and "coordination accuracy", and referring to actual manufacturing level, coordination error sources that need to be adjusted and controlled are taken as the optimized variable. Taking the limit manufacturing level of each component dimension as constraint condition and assembly error requirement as additional constraint condition, firstly the information gain method is used to calculate and adjust the weight of manufacturing cost and coordination accuracy, establishing the target function of tolerance allocation optimization in assembly coordination process synthetically. As a matter of fact, solution of the function is a

Digital Assembly Coordination and Quality Controlling Technology
for Aeronautical Thin-walled Structures

nonlinear programming problem. According to the physical position of the identified coordinating error sources in propagation network, augmented Lagrangian method and penalty function method are often adopted based on models of diagnosing coordinating errors on main network and local sub-network. Then the large-scale nonlinear coordination tolerance allocation problem can be transformed into a series of unconstrained programming sub-problems, with intelligent algorithms, such as ant colony, are used to solve these problems. However, to ensure the optimization convergence, whether the transmission results of each error item in sub-network meet requirements of coordination errors is the key point. After substituting the optimized error links to the nonlinear cumulative calculation process of coordination errors, iterative solution of the sub-problems can be completed. However, more verification works still need to be done. The assembly/coordination quality, which combines assembly deformation information with practical assembly accuracy information, is to be analyzed and evaluated with the results of simulating verification in dimension control software 3DCS. Based on the simulation results, physical state of product and tooling equipment can be adjusted in real-time, and coordination accuracy is evaluated according to latest precision status. In a word, through the continuous improvement of assembly coordination process, the active control of assembly consistency can be realized.

2.3.2 Working Mode for the Whole Manufacturing Stage Based on Digital Coordination Model

Aircraft manufacturing is a progressive assembly process of parts, components, and the whole aircraft, which are assembled using different kinds of assembly tooling, jigs, and stations. The process is generally divided into four portions: parts manufacturing, module unit assembly, component assembly, and the jointing of large components, each has an obvious hierarchical characteristic. Therefore, according to the organization hierarchy relationship and the detailed application objects of aircraft, to make DCM available for all manufacturing stages, the classification of DCM is shown in Figure 2-11.

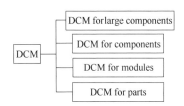

Figure 2-11　Classification of DCM

Based on the composed detailed information in DCM and the classification results, a coordination route that containing the geometric error/size/shape/position transferring process is proposed, and working mode for the whole manufacturing stage based on DCM can be shown in Figure 2-12. Combining with the designed MBD model of products, and under the new normal of digital manufacturing environment, DCM is taken as the basis for coordination process in assembly. It acts the sole coordination data source

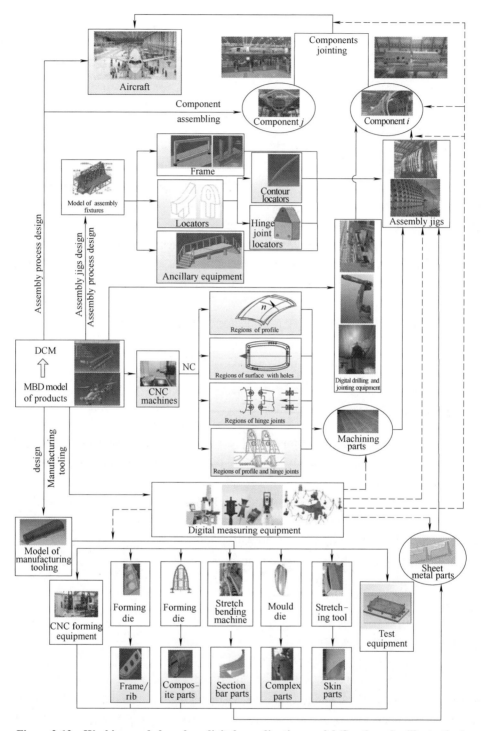

Figure 2-12　Working mode based on digital coordination model (See the color illustration)

for ① forming tooling design, ② CNC machining, ③ assembly tooling/jigs design and installation, ④ assembly process design, and ⑤ positioning/jointing/measuring operations for the whole assembly process. The one-way arrow indicates the applicable scope of DCMs for ① sheet metal forming, ② NC machining, ③ drilling and jointing, ④ assembling process. The arrows also mean the flow process of dimension/shape/spatial positions from the designed aircraft model to the final manufactured aircraft from one original data. The dotted arrow represents the detection and inspection operations on products and their corresponding manufacturing tooling with relevant digital measuring equipment.

From the above analysis, in this new working mode based on DCM, it is shown that the foundation of this mode is DCM, the focus is error transfer process in assembly, the direct analysis basis is measuring data, the core factor is dynamic feedback adjustment, and the ultimate goal is to control the consistency of accumulated errors between assemblies. In constrast to the working mode based on hard master tooling, the function process of this mode is a closed loop that CEs and CRs can be adjusted based on real-time digital measuring results. The detailed function process of DCM in the whole aircraft manufacturing can be shown in Figure 2-13. It is mentioned that only main part types and frequently used measuring equipment, such as coordinate measuring machine, laser tracker, force sensor, displacement sensor, and indoor GPS, are listed here.

When coordination accuracy doesn't meet the requirements specified in DCM, a dynamic feedback adjustment for the whole manufacturing process should be done as necessary, and two main solutions are proposed here to achieve the goal, as shown in Figure 2-14.

Step 1 When parts or components haven't been put in practical assembly scene, a measuring field is built firstly with different kinds of measuring equipment. With the help of a normalized conversion interface, different types of measuring data can be integrated together. Then, ① practical geometry error items of the CEs for parts and tooling, ② spatial position status of parts and tooling, and ③ information of assembly stress and assembly deformation can be gained in the field. Based on these real-time measurement results, after a lot of operations such as registration, fitting, denoising, lightweight processing, a temporary process model DCM' would be reconstituted to modify the theory DCM. Unlike the designed product MBD model that has a static description of error tolerance, this new model is dynamical and keeps in accordance with the practical manufacturing conditions. The above actions can provide a model foundation, which derived from the live manufacturing event, for the subsequent optimization of different assembly behaviors. Based of practical measuring data, some improvement actions could be taken, such as ① optimizing assembly sequence, ② optimizating partitions for the

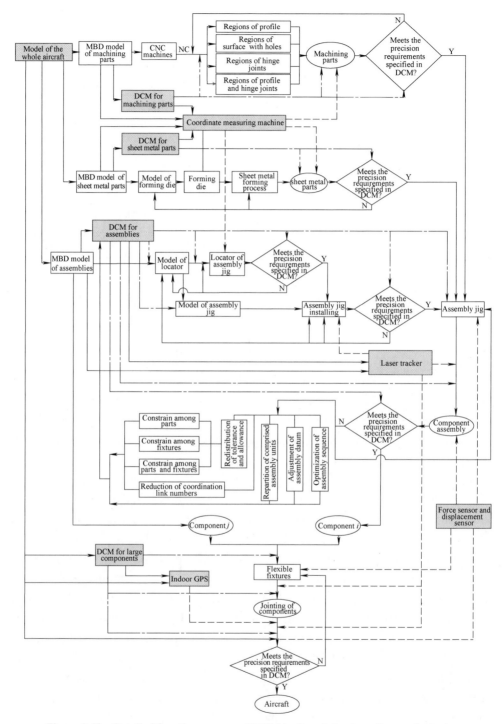

Figure 2-13   Detailed function process of DCM in the whole aircraft manufacturing

Digital Assembly Coordination and Quality Controlling Technology
for Aeronautical Thin-walled Structures

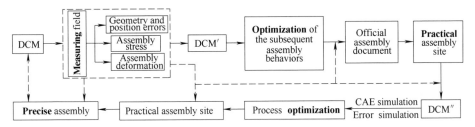

Figure 2-14   Dynamic feedback adjustment for manufacturing process

products and assembly modules, ③ simulating and optimizing assembly precision based on relay calculations that consider actual physical deformation and geometric variation, ④ modeling the correlation relationship of positioning accuracy for different end locators, ⑤ unified planning of reference datum systems at different manufacturing stages, ⑥ reallocation of tolerances and allowances, ⑦ reducing the number of coordination error links, and others, as donated by the dash-dot lines in Figure 2-9. With an adjustment of the status for product and tooling, another measuring work is done to gain the current assembly status. When assembly accuracy meets the required precision, the official assembly document can be compiled and put into the practical assembly site for guiding the engineering operations.

    Step 2   When parts or components are put in practical assembly scene, an iterative process of simulation and optimization for CEs and CRs is done. The assembly physical deformation, caused by ① manipulating forces in assembly process, and ② up and down operations for aircraft parts or components with assembly jigs, would influence the coordination accuracy seriously. The above kinds of error items, such as variation of spatial position, displacement, angle, stress, and others, can be measured and tracked in real-time to acquire practical detailed numerical value. For a given assembly step, if the assembly accuracy cannot meet the requirements, assembly operations should be stopped temporarily. The perception operation should be repeated, and the reconstituted model that mentioned in Step 1 is to be updated, and can be donated as DCM″. By combining deformation error with certain CAE systems and error simulation systems, such as 3DCS, another simulation and optimization actions should be done, and the adjusted results would be delivered to the practical assembly site for starting the temporarily suspended assembly operations. Lastly, a more accurate assembly and coordination precision can be gained.

2.4   Experimental Verification

    There are four wing flap components in a certain type aircraft: the inboard component, the outboard component and two additional symmetrical components. The main parts are ribs, spar(s), hinge joints, skin panel, as shown in Figure 2-15. The two components would be joined together at the interface of the 14th rib and the 15th rib to form a single component. At the macroscopic level, the position and number of ribs that distributed on the spar are different from the inner and the outer components, with the length of the two assemblies differing by about 300mm.

Verification of the working mode is carried out based on flexible assembly process for the four components, which means they are assembled on the same assembly fixture. Thus, position of the end locating effector in this fixture should be adjusted to fit the different components. And a set of locating units with three motion axes should be designed to meet this requirement (Figure 2-15). To assemble the components in digital manufacturing environment, hard master tooling would not be used in the manufacturing process; instead, DCM is taken as manufacture basis for ① design/fabrication of assembly fixtures, and ② assembly of the components. In designing the assembly process, firstly, assembly datum of the components is selected based on profile of skeleton parts, and the detailed assembly procedure can be shown in Figure 2-16. Contrasting to locating method based on contour boards, coordination holes that based on DA technology[11,27] are used for locating the ribs, as shown in Figure 2-17.

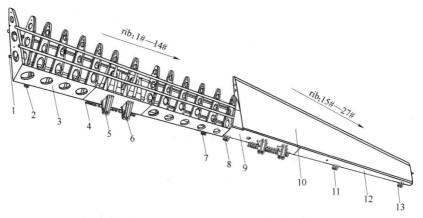

Figure 2-15　Structure of the wing flap components

1—stringer; 2/4/7/8/11/13—hinge joints on the spar; 3/9/12—spar;
5/6—manipulating joint on the 7th and 8th rib; 10—skin

Locating spar	Locating rib	Riveting rib and spar	Locating stringer	Riveting rib and stringer	Covering skin panel	Compacting and riveting skin with skeleton parts	Measuring assembly error	Completed

Figure 2-16　Wing assembly process based on skeleton parts

According to transfer process of coordination datum, combing with engineering data that required in flexible assembly process, process datum reference system is constructed

Digital Assembly Coordination and Quality Controlling Technology for Aeronautical Thin-walled Structures

firstly based on 3D mathematical MBD model of products. Next, the product CEs and their corresponding machining/locating/jointing/measuring parameters are defined according to CMCA. Taking ① design and manufacturing datum system, and ② assembly precision requirement into account, CR information is analyzed on the basis of key coordination characteristics. Then, the DCM of the right inboard component is built under CAA for CATIA environment, as shown in Figure 2-18, and detailed contents that comprised in DCM are shown in Table 2-1.

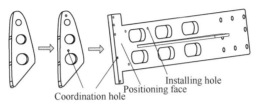

Figure 2-17   Locating method based on coordination holes

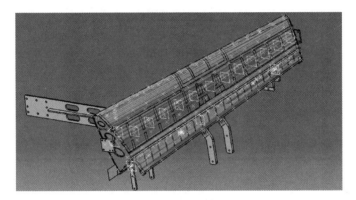

Figure 2-18   DCM of the right inboard component

Table 2-1   Detailed contents in DCM

DCM		Comprised contents
CE	Coordinate system	Product design coordinate system; fixture coordinate system; measuring coordinate system
	Datum element	Suspension plane 1—3; rib plane 1—14; horizontal datum plane
		Rotation axis of the hinge joints
		Theory profile surface of skin and ribs
	CE of component	Coordination holes on ribs; profile surface of ribs; inner and outer profile surface of skin; web side of ribs/spars; holes of hinge joint
	CE of flexible fixture	Fixture reference plane; position of coordination hole on rib locators at different assembly stations; web side of locator on different locating units; surface for locating spars and hinge joints
	CE of machining process	Detailed geometric shape; machining/forming methods and parameters
	CE of measuring process	Measuring points on the surfaces of rib, skin, spar and fixture; measuring points on reference datum axes; measuring points on locating surface

DCM		Comprised contents
CR	Coplanarity and normal angle between planes/surfaces; colinearity between lines/curves; angle/parallel/ tangent constraints	Mating and positioning precision relations between ribs and their relative locating units; manufacturing precision relations between ribs and their relative forming dies; aerodynamic profile precision relations between ribs or other skeleton parts; corresponding precision relations between forming die of ribs; precision relations between coordination holes that distributing on locating board

With the use of DCM, coordination design process for geometric size/shape and tolerance/allowance information of the parts/components/fixtures is put into action. The above information is directly delivered to the manufacturing equipment and the practical assembly site as digital data. Lastly, with the help of CNC system of flexible locating units, a measuring field is built with different kinds of measuring equipment under the guidance of CMCA, allowing measuring operations of parts/fixtures/assemblies and assembling work of wing flap components. It is mentioned that the detailed geometric relationship would be modified and reconstructed according to the measured and analyzed data to keep in accordance with practical assembly site.

Synthesizing the above analysis, according to the working mode based on DCA that illustrated in Figure 2-12 and the structure of components/end-locators, coordination route, which includes the transfer process of error items and digital geometry dimension/shape values in the whole manufacturing process, is planned, as shown in Figure 2-19.

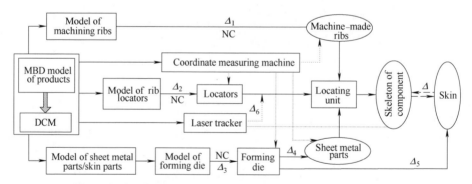

Figure 2-19   Coordination route of the component based on DCM

In the experiment that described in this chapter, coordination process based on DCM is divided into three stages: ① coordination process for designing flexible assembly fixtures; ② coordination process for manufacturing parts of the component and the fixture; and ③ coordination process for fixture installation and components assembly. The detailed function process can be described as follows.

2.4.1 Coordination Process for Designing Flexible Assembly Fixtures

Based on the analysis of the products to be assembled, CEs of the wing flap component are the key coordination features that defined in DCM. With these features, by quoting ① the datum elements, ② geometric shapes/sizes, and ③ their tolerance/allowance information that contained in DCM, the design of flexible assembly fixtures can be brought into action.

With the reference to datum system defined in DCM, the direction of the fixture's coordinate system is determined firstly. The direction of X axis is established by the line between centers of hinge joints on spar that are distributed across four components. The direction of Y axis is taken as the normal vector of datum plane of the wing. And Z axis is parallel to datum plane of the wing. Then with reference to key locating features and geometric datum planes of the wing and ribs, the space placement form of four components is finally determined in the fixture's coordinate system. It is mentioned that, angles of above three axes would be modified according to the calculations of the establishment error on practical assembly site with certain measuring equipment.

Secondly, three types of end locators are designed, as shown in Figure 2-20. The first type is locators for ribs, using locating method of "one plane with two holes". The design basis is web plane of the rib and coordination holes that defined on ribs in DCM. It is mentioned that in designing process of the special rib locators (Figure 2-13), a function of locating a total of 46 ribs of the four components, excluding the 8 end ribs, should be realized. The locating board is designed depending on the size and position relationships of the 46 ribs. By quoting ① the datum planes of wing, and ② the web plane of ribs that defined in DCM, a new process plane is created. Positions of coordination holes are located at the intersection line of the above new plane and the ribplane. Then, the ribs are divided into 4 groups, with the consideration of size of different ribs and locating space/stability. A group of four holes are designed on the locators, corresponding to the four divided groups of ribs. The two web-side surfaces of locating board would have a function of locating the web sides of ribs. Another common locating hole lies at the bottom of the board. What's more, the bottom hole, together with the four grouped

Figure 2-20   The designed locators

locating holes, would also be taken as the OTP (optical tooling points) points in practical installation process of the fixture and the assembly process of ribs. When their positions meet the accuracy requirements, they can be secured for locating.

The other two type locators are used for locating spar and hinge joints. In the same way, the design basis includes ① web of spars, and ② axis and suspension plane of hinge joints, which are also already defined in DCM.

Then, according to ① the size and position differences of four components, and ② the grouped product's CEs that to be assembled on each locating unit, general layout relationship between seven locating units (with the 7th locating unit locates at the opposite side of the 2nd and 4th locating units, which is not shown here) and locating travel for all moving directions can be optimized, taking the assembling space and economic factors into account. When the flexible assembly scheme is successfully simulated and verified in DELMIA environment, the general layout relation of the seven locating units can be eventually determined, as shown in Figure 2-21.

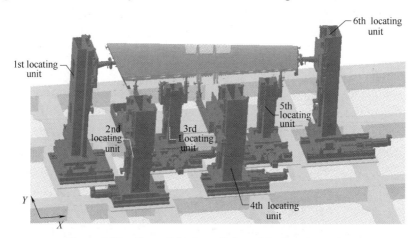

Figure 2-21   General layout relation of the locating units

2.4.2   Coordination Process for Manufacturing Parts of the Component and Fixture

In the manipulating process of products and fixture parts according to the independent manufacturing principles, the datums that defined in the DCM, such as the axes of coordination holes, and the locating surfaces, should be changed to the actual manufacturing datums. Taking the manufacturing method of NC machining as an example, after the machining process and post processing process are designed, the geometric and tolerance information would be delivered to CNC machines through NC orders with a digital medium from DCM. In summary, the machined parts can be regarded as the materialized digital data. In the machining process, strict quality schemes should be developed for controlling the key process parameters, such as ① spindle speed, ② cutter feeding rate, and ③ cutting dosages. When the measured error results meet the required accuracy that contained in DCM, the parts can then be used for further installation and assembly operations. Further detailed description will be shown in Chapter 4 and Chapter 5.

2.4.3 Coordination Process for Practical Fixture Installing and Components Assembling

At the coordination stage of installing the fixture, according to the datum elements, and the theory data of measuring points that defined in DCM, coordination process for practical fixture installation and components assembly is carried out. With the guidance of the control method for coordination accuracy, the required assembly precisions of the hinge joints are gained, as shown in Table 2-2.

Table 2-2　Assembly precisions of the hinge joints for the inner component

Hinge joint	2#	4#	5#	6#
Precision/mm	−0.071	0.042	−0.019	−0.058

For the inner wing flap component, the theory optimized assembly errors of skin profile at the comprised fourteen ribs can be shown in Figure 2-22, which is colored in yellow.

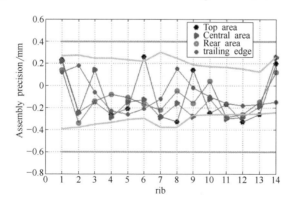

Figure 2-22　Comparison of skin profile precision for the inner component
(See the color illustration)

Then, the practical measured assembly precision at the top/central/rear/trailing-edge areas of the skin surface is also shown in Figure 2-22, which is colored in blue. In addition, the required assembly accuracy that defined in official assembly document is colored in green, with a range of [+0.40, −0.60] mm. Based on the analysis of these actual assembly data, two conclusions can be drawn: ① the actual assembly precision fits well with the calculated theory precision under the guidance of CMCA contained in DCM, and ② the assembly accuracy can successfully meet the design requirements.

However, for the whole wing flap component, the flush coordination accuracy between skins of the inner and the outer components is also required to guarantee in official assembly document. Where flush is the offset or altitude difference between two different surfaces. The red portion that marked in the figure is coordination region, and the direction of flush is perpendicular to skin surface of the ending ribs in this case. With the help of CMCA, undoubtedly, the consistency of accumulated assembly errors, both in the error value size and direction, can be controlled at a good level in theory. However, due to the measuring field was built with laser tracker in our experiment, only certain key points on the surface could be measured. In this situation, error items with maximum difference for the inner and outer components are recorded in Table 2-3. Where ① "R" represents the right component and ② "L" stands for the left one. However, for the older wing flap components, only the flush resulted within the working mode based on hard master tooling can be collected. But to the working mode based on DCM, the average assembly precision of the right and left components on this flexible assembly fixture is put forward. The conclusion that there is an enhancement of about 30% in the coordination accuracy can be drawn by comparing these two working modes.

Table 2-3　The measured practical profile accuracy

Assembly precision	Inner component	Outer component	Flush
Working mode based on hard master tooling/mm	—	—	0.840
Working mode based on digital coordination model/mm	0.262 (R)	−0.327 (R)	0.589
	−0.304 (L)	0.214 (L)	0.528
Coordination accuracy enhancement/%			≈30

To be more convinced, a comparison of efficiency enhancements between the working mode based on hard tooling and the working mode based on DCM is carried out. The assembly periods of the comprised components based on the two working modes are shown in Table 2-4. It can be found that the efficiency of the working mode based on DCM has increased about 40%. Together with Table 2-4, the conclusion that a higher assembly quality and an enhancement in aircraft's flight performance could be drawn under the working mode based on the proposed DCM in this paper.

Table 2-4　Comparison of assembly period for the two working mode

Assembly period	Inner component	Outer component
Working mode based on hard master tooling/d	12	10
Working mode based on digital coordination model/d	7.5	6
Efficiency enhancement/%	37.5	40

Digital Assembly Coordination and Quality Controlling Technology for Aeronautical Thin-walled Structures

2.5 Summary

① The defined digital coordination model is a process model that mirrors practical assembly site, consisting of coordination element, coordination relationship, and control method for coordination accuracy. It can replace the hard master tooling and act as the only basis for the transfer process of dimension/shape/position in aircraft manufacturing.

② The principle of the working mode based on digital coordination model is proposed, and the detailed function process of DCM in the whole aircraft manufacturing stage is presented. It is shown that the foundation of the mode is DCM; the focus of the mode is error transfer process in assembly; the direct analysis basis of the mode is measurement data; the core factor of the mode is the dynamic feedback adjustment; and the ultimate goal is to control the consistency of accumulated errors between assemblies. Contrasting to the working mode based on hard master tooling, under the measurement field and virtual simulation/analysis environment, the function process of this mode is a dynamic closed loop in which the CEs and the CRs can be adjusted in real-time continuonsly with CMCA.

③ Flexible assembly process of four wing flap components is taken as the experiment object, and the corresponding DCM is built for the practical manufacturing process. Beneficial results showed that the propagation of dimension/shape/position for the product and the flexible tooling can be completed under the new working mode based on DCM. Further more, assembly accuracy can fit the calculated results and the design requirements well, and efficiency of the working mode based on DCM has increased about 40%.

④ This new coordination working method can improve assembly quality and efficiency in practical assembly process undoubtedly, but its needs a combined help of 3D design technology, information technology, measuring and control technology, and other advanced manufacturing technologies, making its systemic and universal application very difficult. Thus, the application scope of digital coordination mode should be overall planned to fit the practical manufacturing situations, especially for the mixed using of the two working modes. To eliminate the unnecessary error links, there should be an unambiguous definition of which working mode should be employed, and the coordination basis should be the unified for complex components. Balancing the mingling relationship between the two working modes is an important subject that deserve to be discussed at the current manufacturing stage.

⑤ Aircraft manufacturing quality is the main focus for the designers and manufacturers. To ensure the consistency of accumulated assembly errors between assemblies, minimizing locating and clamping deformation for flexible parts, managing error transfer efforts as the assembly process flowing, and controlling method of residual stress during assembly process, should be enhanced as well.

References

[1] Bullen G. Automated/Mechanized Drilling and Countersinking of Airframes[M]. Warrendale: SAE International Press, 2013.

[2] Williams G, Chalupa E, Billieu R, et al. Gaugeless tooling[R]. SAE Technical Paper, 1998.

[3] Gitta M. Destination. Myachkovo Airport[M]. Los Angeles: Sent Publishing, 2013.

[4] Manohar K, Hogan T, Buttrick J, et al. Predicting shim gaps in aircraft assembly with machine learning and sparse sensing[J]. Journal of Manufacturing Systems, 2018, 48: 87-95.

[5] Andolfatto L, Thiébaut F, Lartigue C, et al. Quality- and cost-driven assembly technique selection and geometrical tolerance allocation for mechanical structure assembly[J]. Journal of Manufacturing Systems, 2014, 33(1): 103-115.

[6] Han X, Li R, Wang J, et al. Identification of key design characteristics for complex product adaptive design[J]. International Journal of Advanced Manufacturing Technology, 2018, 95(1-4): 1215-1231.

[7] Campos A, Johnson R, Kennedy R. System and method for wiring an aircraft[P]. US14457514, 2017-08-08.

[8] Andrés P, Oliva M, Racero J, et al. Design and Implementation of a Prototype for Information Exchange in Digital Manufacturing Processes in Aerospace Industry[J]. Product lifecycle management and the industry of the future, 2017, 517: 590-600.

[9] Ivanov I, Kryukov I, Larina E, et al. Mathematical and software support for 3D mathematical modelling of the airflow impact on the optical-mechanical unit mounted in the aircraft unpressurized compartment[J]. Moscow Aviation Institute, 2017, 10(4): 113-123.

[10] Quintana V, Rivest L, Pellerin R, et al. Will model-based definition replace engineering drawings throughout the product lifecycle? a global perspective from aerospace industry[J]. Computers in Industry, 2010, 61(5): 497-508.

[11] Lucas A, Irving M, et al. Implementing determinate assembly for the leading edge sub-assembly of aircraft wing manufacture[J]. SAE International Journal of Aerospace, 2014, 7(2): 246-254.

[12] Genta G, Galetto M, Franceschini F. Product complexity and design of inspection strategies for assembly manufacturing processes[J]. International Journal of Production Research, 2018, 12: 1-11.

[13] Chen Z, Du F, Tang X, et al. A framework of measurement assisted assembly for wing-fuselage alignment based on key measurement characteristics[J]. Inderscience Publishers (IEL), 2015, 10(2): 107-128.

[14] Li X, Dang X, Xie B, et al. Flexible tooling design technology for aircraft fuselage panel component pre-assembly[J]. Assembly Automation, 2015, 35(2): 166-171.

[15] Bolotin L, Johnson S. Radio frequency identification system for use with an assembly

Digital Assembly Coordination and Quality Controlling Technology
for Aeronautical Thin-walled Structures

line[P]. US201514746343, 2016-08-23.

[16] Dner R, Broll W, Grimm P, et al. Virtual reality und augmented reality (VR/AR)[J]. Informatik-Spektrum, 2016, 39(1): 30-37.

[17] Grieves M, Vickers J. Digital twin: Mitigating unpredictable, undesirable emergent behavior in complex systems[J]. Springer International Publishing, 2017, 17: 85-113.

[18] Wijaya T, Caesarendra W, Pappachan B, et al. Robot control and decision making through real-time sensors monitoring and analysis for industry 4.0 implementation on aerospace component manufacturing[C]. IEEE Pacific Rim Conference on Communications, Computers and Signal Processing, Columbia, 2017.

[19] Kellegoz T. Assembly line balancing problems with multi-manned stations: A new mathematical formulation and Gantt based heuristic method[J]. Annals of Operations Research, 2017, 253(1): 377-404.

[20] Rashid A, Masood T, Erkoyuncu J, et al. Enterprise systems' life cycle in pursuit of resilient smart factory for emerging aircraft industry: a synthesis of critical success factors'(CSFs), theory, knowledge gaps, and implications[J]. Enterprise Information Systems, 2018, 12(1-5):96-136.

[21] Wang Q, Zheng F, Ren Y, et al. Posture evaluation method for aircraft component based on hole feature[J]. Computer Integrated Manufacturing Systems, 2017, 23(2): 243-252.

[22] Yang B, Wang Z, Yang Y, et al. Optimum fixture locating layout for sheet metal part by integrating kriging with cuckoo search algorithm[J]. International Journal of Advanced Manufacturing Technology, 2017, 91(1-4): 327-340.

[23] Frumuǎ A, Epureanu A. Part accuracy management by topological mapping of deviations[J]. Applied Mechanics & Materials, 2017, 859: 210-216.

[24] Rebello A, Ostrowski M, Yokoyama K, et al. Method and system for creating a tooling master model for manufacturing parts[P]. US20020683699. 2003-08-07.

[25] Boeing Next-Generation 737[EB/OL]. (2024-12-03)[2024-12-03] http://www.boeing.com/commercial/737ng/.html

[26] Xi F, Lin Y, Li Y. A robotic percussive riveting system for aircraft assembly automation[J]. Springer International Publishing, 2017, 19: 443-468.

[27] Guo F, Wang Z, Kang Y, et al. Positioning method and assembly precision for aircraft wing skin[J]. Proceedings of the Institution of Mechanical Engineers Part B Journal of Engineering Manufacture, 2016, 232(2): 317-327.

Chapter 3
Comprehensive Identification of Coordination Features with Complete Importance Modeling

3.1 Introduction and Related Work

Coordination feature (CF) is the coordination information carrier of aircraft product and is used to transfer the information of size and shape during aircraft manufacturing. The consistency of accumulated assembly errors between two or more assemblies, namely coordination accuracy, is affected by the changes of CF's geometric size, shape, spatial position, and other attributes. At the same time, these features are usually the focal adjustment and control portion when coordination accuracy cannot meet design requirements. To get an accurate coordination precision, hard master tooling is taken as the manufacturing basis in aviation products for a long time[1]. The CFs of product often have a corresponding and matching relationships, i.e. the geometric or spatial position constrains, with the CFs that are limited in the field of hard master items, such as measuring templates, prototype workpieces, and standard metric gauges. Under this working method, a qualified manufacturing precision and assembly process of the product's CFs, as well as the coordination accuracy between different assemblies, would fit the design requirements. However, due to the tedious dimension transfer links originating from these master entities, coordination accuracy is often difficult to guarantee. With the development of digital manufacturing environment nowadays[2-5], this kind of working mode is rarely used for economic and precision reasons, leading to a lack of the direct manufacturing basis in dimension and shape transfer processes for products. Because of the lack of hard master tooling, coordination accuracy has a tight relationship with the whole detailed assembly process of different kinds of parts or components. Under the digital manufacturing environment, the traditional experience based on hard master tooling is not suitable anymore for identifying the CFs of product and enhancing

Digital Assembly Coordination and Quality Controlling Technology
for Aeronautical Thin-walled Structures

the coordination accuracy. In addition, due to the complexity of aircraft structures, the whole manufacturing process can be divided into multiple assembly stations or manufacturing stages. As the manufacturing process flows, assemblies of lower hierarchies often act as special parts for higher assembly hierarchies[6,7]. Taking this complex interaction and evaluation relationship between assemblies into account, CFs of a certain components at different manufacturing stage may differ from each other. In summary, for aircraft products, to guarantee coordination accuracy in the multi-hierarchy and multi-station assembly processes, it is meaningful to identity the CFs in digital manufacturing environment.

Based on the signification mentioned above, researches have paid much attention to CF identification[8,9]. Two steps are generally contained in the identification process: the qualitative analysis and the quantitative analysis. In the first step, with the consideration of ① structure and function of the product, ② coordination accuracy requirement, and ③ engineering experience, among other foctors, influencing factor sets $\{c_1, c_2, \cdots, c_n\}$ that affecting the CFs $\{y_1, y_2, \cdots, y_m\}$ at different assembly hierarchies is determined consequently. Three methods are frequently used: the method based on historical data and knowledge of the similar products, the method of expert adjusting (Deferr method), and the risk analysis method[10]. These methods have certain applicability for products with simple structures, but for the complex products, the prioritized results may not accurate enough, leading to redundant selections. In the next step, the CF is further optimized from the alternative feature sets in Step 1. Mathieu[11] adopted an assembly-oriented graph (AOG) methodology to identify the CFs according to the product's design and manufacturing information. Andolfatto[12] considered that KCs could be expressed as geometrical conditions, even for any measurable characteristic, and performed a tolerance allocation based on datum flow chain (DFC) to ensure the product's KCs kept within the authorized boundaries. However, the identification efficiency of the above two methods is not high enough, and the analysis process should be repeated if the assembly objects are changed. To solve this problem, Johansson[13] described an identification method by calculating the risk priorities, using variation mode and effect analysis (VMEA). Tang[14] developed a quantitative analysis method takes into account of the Taguchi quality loss function and the loss sensitivity. And different mathematical models were proposed to determine the relative importance of the comprised factors, providing helpful guidance of the quantitative identification for aircraft product CFs[15-19]. Aiming to secure a product's geometry quality, Jareteg[20] gathered all activities that related to a product's variation in manufacturing and calculated its insensitivity to variation during its development. To present the data and relationship between weld geometry characteristics/ parameters and influencing characteristics, Karlsson[21] developed a matrix flow chart and interrelationship matrix, offering a generalized and easy way. In addition, Madrid[22] systematically described how a product's performance was affected by the variation that occurs throughout its own fabrication process, then identified the factors or parameters controlling the transformation process of product characteristics, and lastly determined the causes of variation. Generally speaking, the whole aforementioned identification methods could provide important reference for identifying aircraft coordination feature. However, the situation where hard master tooling is no longer used in digital manufacturing environment is not considered specially. Furthermore, the identified results may not have a close relationship with coordination accuracy. In addition, the conformity for practical production situation and the characteristics of multi-hierarchy process for aircraft products deserve discussion. Ma[23] firstly selected 22 quality characteristics from 60 normal samples and 30 abnormal samples, then cut them to the final 16 using Relief algorithm. However, the number of options is still too large, causing difficulties in actual production. What's more, the elimination parameters is subjectively determined, and how to set up the reasonable weight and how to further adjust the filter criteria should be further explored. To help designers in rapidly developing a design scheme, Han[24] identified the KCs resulting in some functions, empirical formulas and evolution rules. But only the design factors were considered, and a systematic method of identification to support product life cycle need to be developed.

In this chapter, a comprehensive method for identifying CFs of aircraft product based on importance degree calculation is developed, and its engineering application is presented. Firstly, the interaction and evaluation relationships between assemblies at different manufacturing stages are analyzed. Secondly, the quality loss between two adjacent assembly hierarchies is calculated. Thirdly, to eventually identity the CFs for aircraft products, the concepts of centrality degree index and cause degree index are proposed for calculating the complete importance degree. Fourthly, two practical engineering cases are presented to indicate the methodology's feasibility, followed by some conclusions. In the final section, we give some concluding remarks.

3.2 Hierarchical Decomposition of Coordination Features for Aeronautical Product

Due to the characteristics of multi-hierarchy process for aircraft products and multi-station for assembly process, the decomposition of product CFs is closely related to the manufacturing process. According to the three factors listed below, namely, the product structure, the coordination working method, and the coordination accuracy

Digital Assembly Coordination and Quality Controlling Technology for Aeronautical Thin-walled Structures

requirements. Firstly, the product structure tree is built in a hierarchical form. Classification of the CFs from the top hierarchy can be shown in Figure 3-1, which shows the comprised components/modules/parts in a multi-stage form.

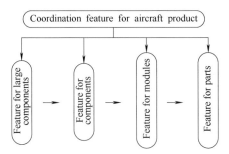

Figure 3-1   Classification of coordination features

When assembling an aircraft in digital environment (i.e., when the hard master tooling in manufacturing is not used anymore), with the reference of the divided breakdown interface and other manufacturing information, the manufacturing tree is built taking the coordination accuracy requirements between assemblies into account. It is mentioned that the manufacturing tree have a difference with product structure tree, which can reflect the assembly process for different kinds of parts or components. Then, the CFs of top hierarchy for aircraft product would be gradually decomposed along the manufacturing tree. And the interaction and evaluation relationships of the decomposed CFs, which are distributed on different assemblies and diverse manufacturing stages, are analyzed, as shown in Figure 3-2.

Where the dotted lines with double-headed arrows represent the coordination accuracy requirements at different manufacturing stages; and the CFs, which require more attention in their manufacturing process, are marked with the "KEY" symbol. The hierarchical form of the decomposed CFs can be explained with the following process in detail.

Hierarchy 1   CFs for aircraft product, such as the skin profile and the horizontal measuring points of the wing. For the whole aircraft, the top CF, which is influenced by a number of CFs distributed on different components/modules/parts, is donated as y_0.

Hierarchy 2   CFs for components. To get an accurate profile, many kinds of components or large components should be jointed together at certain design/process interfaces or hinge-joint areas. The relationship between these components' CFs and the top CF can be modeled by Equation (3-1).

$$y_0 \xrightarrow{\text{decomposed}} \text{SubCF}(x_{1,m_1}^1, x_{2,m_2}^1, \cdots, x_{n_1,m_{n1}}^1) \tag{3-1}$$

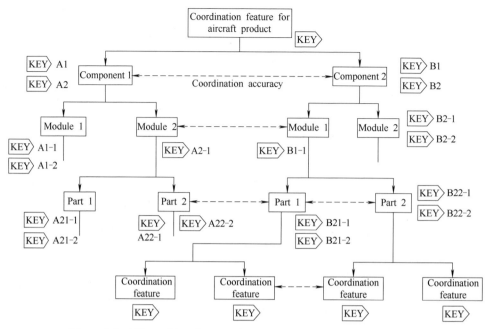

Figure 3-2   Hierarchical form of the decomposed coordination features

Where, the symbol $x_{i,j}^k$ represents the comprised CFs; k represents the assembly hierarchies, i.e., the components assembly, the modules assembly, and the lowest parts, with corresponding values of 1, 2, $\cdots$, n; i indicates one certain specific comprised components/modules/parts in the present assembly hierarchy; j represents the number of CFs at each assembly hierarchy; and the symbol "$\rightarrow$" reflects the interaction and evaluation relationships of the relevant CFs between two adjacent assembly levels.

Hierarchy 3   CFs for modules. Modules are the comprised elements for component and are assembled by different parts at a lower hierarchy. To guarantee the coordination accuracy between different components or between different components and modules, CFs of different modules at the current assembly hierarchy should be identified. The decomposition process for components can be expressed by Equation (3-2), where p, q, and s represent the number of the decomposed CFs for different parts.

$$
\begin{aligned}
x_{1,m_1}^1 &\xrightarrow{\text{decomposed}} \text{SubCF}(x_{1,m_{11}}^2, x_{1,m_{12}}^2, \cdots, x_{1,m_{1p}}^2) \\
x_{2,m_2}^1 &\xrightarrow{\text{decomposed}} \text{SubCF}(x_{2,m_{21}}^2, x_{2,m_{22}}^2, \cdots, x_{2,m_{2p}}^2) \\
&\ \ \vdots \\
x_{n_1,m_{n1}}^1 &\xrightarrow{\text{decomposed}} \text{SubCF}(x_{n_1,m_{n11}}^2, x_{n_1,m_{n12}}^2, \cdots, x_{n_1,m_{n1s}}^2)
\end{aligned}
\tag{3-2}
$$

Hierarchy 4 CFs for parts. For a part located at the lowest assembly hierarchy, the comprised CFs are represented here by a symbol of $x^3_{u_i,v_i}$. Where v_i stands for the CF numbers on part u_i. And the precision of these parts' CFs is the foundation of guaranteeing the consistency of accumulated assembly errors between different modules/components.

3.3 General Framework of the Comprehensive Identification of Coordination Features

Based on the analysis in Section one, it is known that the identification of CFs in digital manufacturing environment is a meaningful work, especially in the multi-hierarchy and multi-station assembly process for aircraft products. To identity these "KEY" items, which having a close relationship to assembly coordination accuracy, a comprehensive method based on importance degree modeling is developed, as shown in Figure 3-3.

The general framework of the comprehensive identification method contains two key steps. In the first step, with the consideration of ① product design MBD (model based definition) model, ② coordination accuracy requirements, and ③ the detailed assembly process that shown in the manufacturing tree, the common coordination regions of the product are determined by qualitative analysis. Then based on ① the mating features at three adjacent assembly hierarchies, i, j and k, ② the manufacturing hierarchical relationship of the decomposed product CFs, and ③ the manufacturing error relationship, two variation propagation chains (VPCs) of the assemblies across the three assembly hierarchies are constructed. The VPC, which is constructed based on the mating constrains of each CF, is an error accumulation route of the CFs' variation in the assemblies. The relationship between the error variation of alternative CFs and the assembly coordination accuracy can be reflected by a special VPC. In the VPC, the start feature, the end feature, and the intermediate features with dimensional or mating relationships are included in detail. When finishing the steps mentioned above, the alternative CF sets $\{c_1, c_2, \cdots, c_n\}$ and the target CF sets $\{y_1, y_2, \cdots, y_m\}$ are determined. In the second step, by combining the Taguchi quality loss function with accuracy principal, the error correction coefficient H is used to calculate the quality loss of the target CFs. As a matter of fact, quality loss is caused by the precision variation of the alternative CFs at different assembly hierarchies. Then, the influence degree and the affected degree of alternative CFs at the jth assembly hierarchy are calculated with the DEMATEL (decision making trial and evaluation laboratory) method[25]. Meanwhile, the

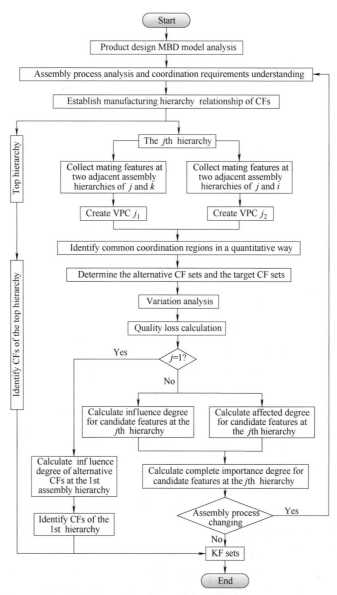

Figure 3-3   General framework of the comprehensive identification of coordination features

concepts of centrality degree index and cause degree index are proposed for calculating the importance degree of alternative CFs at the jth assembly hierarchy. With the above two degree indexes, an algorithm for the calculation of complete importance degree is presented. Then by comparing the values of importance degrees for alternative CFs, the targeted CFs can be eventually identified. In addition, two important attributes of the identified CFs can be judged: whether they are influencing factors or outcome-oriented factors, which would be very helpful in assembly process designing and practical engineering applications. However, for some special situations, for example, ① the top CFs is to be analyzed in Section 3.4 and Section 3.5; ② CFs for the lowest assembly hierarchy, i.e. the parts hierarchy, and the identification results with an assembly process changing, are to be analyzed in Section 3.5.

3.4　Qualitative Identification of Coordination Features

Actually, these CFs have high manufacturing precision requirement, which would have a close relationship to the motion performance, sealing performance and aerodynamic performance of the aircraft. Meanwhile, the relative spatial position between different comprised components is also pertinent. The fact is that, one certain CF's tolerance variation would affect the precision constrain relationships with other mating CFs that contained in VPC. Considering the aircraft design information in the defined MBD model and the manufacturing requirements, the common coordination regions of aircraft product and their detailed corresponding CFs are shown in Table 3-1.

Table 3-1　Common coordination regions and their corresponding coordination features for aircraft product

Coordination regions	Coordination accuracy requirements	Assembly requirement of CFs
Design interface; regions on interchangeable parts or benchmark parts; process interface with a complex coordination accuracy relationship; the same hinge joint or contour profile that located by a serial assembly jigs; contour profile with a complex surface; regions on parts that each has a manufacturing relationship with corresponding forming die and a coordination accuracy relationship with corresponding assembly tooling, etc	Aerodynamic profile	Waviness and smoothness of contour profile; concave-convex defect at jointing areas (which have rivets, screws and welding points); gaps and flush between skins, etc
	Relative spatial position of components; comprehensive accuracy of position and profile	Dihedral angle, mounting angle, and sweep angle errors of the wing or tail components; gaps and flush between manipulating profile and fixed airfoil; coaxiality between hinge joints; comprehensive accuracy of relative position error and profile deviation, etc
	Relative position of modules and parts that comprised in components	Position requirements for reference datum lines, such as the deviation between practical assembly position and the theoretical position of the beam axis, the wing axis, the bulkhead axis, and the stringer axis; positions of locating surfaces, etc
	Locating/assembly precision among different parts or assemblies	Tolerance of critical dimensions and shapes; mating clearances between parts for non-removable assemblies; mating clearances and coaxiality for butt joints and frame-type hinge joints, etc

It is mentioned that, the features which have relationships with aerodynamic profile can be taken as the top CFs. Although the above qualitative identification results (Table 3-1) have a certain applicable range, to determine the detailed composited CFs for a complex aircraft product with millions of parts, it is not effective and accurate enough anymore. We aim to solve this puzzle in next section.

3.5 Quantitative Identification of Coordination Features by Calculating Complete Importance

In this section, the proposed comprehensive identification method is explained based on alternative CF sets $\{c_1, c_2, \cdots, c_n\}$ and target CF sets $\{y_1, y_2, \cdots, y_m\}$ that were previous determined. With complete importance calculation, CFs in the multi-hierarchy and multi-station assembly processes are eventually identified. The detailed modeling process is given below.

According to Taguchi quality loss function[26], when the variation of potential alternative CF (donated as c_i) deviating from its target value, a quality loss would occur, and the calculation process is shown by Equation (3-3) and (3-4).

$$L = k[(\mu - m)^2 + \sigma^2] = k(\delta^2 + \sigma^2) \tag{3-3}$$

$$k = \frac{A}{\left(\dfrac{TU - TL}{2}\right)^2} = \frac{1}{\left(\dfrac{TU - TL}{2}\right)^2} = \frac{4}{(TU - TL)^2} \tag{3-4}$$

Where L represents the quality loss of c_i; k represents the quality loss constant of c_i; δ indicates the quality offset as the mean value μ of c_i deviating from its target value m; σ^2 stands for the variance of c_i; A represents the quality loss coefficient as there is a variation for c_i, which is taken as 1 in this chapter; TU and TL represent the upper limit and the lower limit of tolerances, respectively.

Generally speaking, the variation of one alternative CF at a lower assembly hierarchy would bring a quality loss influence to multiple features at adjacent higher assembly hierarchies. But in view of global situation, the quality loss of one feature at a higher assembly hierarchy may also have a relationship with multiple CFs at adjacent lower assembly hierarchies. Considering this special one-to-many and many-to-one relationships, to analyze the detailed influence degree between two adjacent assembly hierarchies, a function is defined firstly, as shown in Equation (3-5).

$$y_m = F(c_1, c_2, \cdots, c_n) \tag{3-5}$$

Where y_m represents the CF at a higher assembly hierarchy contrasting with the potential alternative CFs, $c_i(i = 1, 2, \cdots, n)$. Assuming that the partial derivative $\partial F / \partial c_i$ exists, with the use of Taylor expansions, the variation Δy_m of the coordination feature

Digital Assembly Coordination and Quality Controlling Technology
for Aeronautical Thin-walled Structures

y_m that caused by Δc_i can be calculated, as given in Equation (3-6). Where Δc_i represents the variation of alternative CFs c_i at a lower assembly hierarchy.

$$\Delta y_m \approx \Delta_\Sigma = \sum_{i=1}^{n}\left(\frac{\partial F}{\partial c_i}\Delta c_i\right) \tag{3-6}$$

Assuming that ① the partial derivatives $\partial F / \partial c_i\,(i = 1,2,\cdots,n)$ are different constants; ② each c_i is a mutually independent random variable whose mean value is also a constant. Then Equation (3-6) can be regarded as the general expression of a dimension error chain, which has a close relationship with coordination accuracy between different assemblies. These partial derivatives represent the error transfer coefficient for each error link (from C_{11} to C_{1n_1}, and from C_{21} to C_{2n_2}, respectively), as shown in Figure 3-4. Where V_{AB} stands for the coordination error.

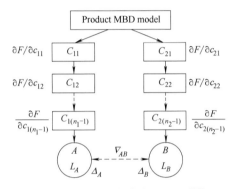

Figure 3-4   Coordination error chain between different assemblies

According to the accuracy principal, then the statistical parameter $((\Delta_\Sigma)_0, \sigma_\Sigma^2)$ of Δ_Σ can be calculated with the reference of the accuracy parameters $((\Delta c_i)_0, \sigma_i^2)$ of Δc_i, as shown in Equation (3-7)~(3-9).

$$(\Delta_\Sigma)_0 = \sum_{i=1}^{n}\left[\frac{\partial F}{\partial c_i}(\Delta c_i)_0\right] \tag{3-7}$$

$$\sigma_\Sigma^2 = H^2 \sum_{i=1}^{n}\left(\frac{\partial F}{\partial c_i}\right)^2 (\sigma_i)^2 \tag{3-8}$$

$$H = 1.8 - 0.8\frac{\sqrt{\displaystyle\sum_{i=1}^{n}\left(\frac{\partial F}{\partial c_i}\right)^2 (\Delta c_i)_0{}^2}}{\displaystyle\sum_{i=1}^{n}\left|\frac{\partial F}{\partial c_i}\Delta c_i\right|} \tag{3-9}$$

In the above three equations [Equation (3-7)~(3-9)], $(\Delta c_i)_0$ stands for midpoint value of the variation Δc_i; H is a correction coefficient of the normal distribution for coordination error calculation during aircraft assembly process. Generally, H is determined according to the practical assembly experience, with a value range of [1, 1.8]. The purpose of using H is to enhance the reliability of the above equations, ensuring the calculation results be more accordant with practical assembly circumstances.

With the reference of the above equations, the quality loss at different assembly hierarchies can be obtained following the next calculation steps.

① Quality loss of the coordination feature y_m at a higher assembly hierarchy, caused by the variation of a single alternative mating feature c_i at the lower assembly hierarchy, can be shown in Equation (3-10).

$$\Delta L_{im} = k_i H^2 \left[\left(\frac{\partial F}{\partial c_i} \right)^2 (\sigma_i'^2 - \sigma_i^2) + \left(\frac{\partial F}{\partial c_i} \right)^2 (\delta_i'^2 - \delta_i^2) \right] \qquad (3\text{-}10)$$

② Quality loss of the coordination feature y_m at a higher assembly hierarchy, caused by the variation of the alternative coordination features $c_1, c_2, \cdots, c_n$ at the lower assembly hierarchy, can be shown in Equation (3-11).

$$\Delta L_m = \sum_{i=1}^{n} \left\{ k_i H^2 \left[\left(\frac{\partial F}{\partial c_i} \right)^2 (\sigma_i'^2 - \sigma_i^2) + \left(\frac{\partial F}{\partial c_i} \right)^2 (\delta_i'^2 - \delta_i^2) \right] \right\} \qquad (3\text{-}11)$$

③ Quality loss of the whole coordination features $Y = \{y_1, y_2, \cdots, y_m\}$ at a higher assembly hierarchy, caused by a single variation of the alternative coordination feature c_i at a lower assembly hierarchy, can be shown in Equation (3-12).

$$\Delta L_i = \sum_{j=1}^{m} \left\{ k_j H^2 \left[\left(\frac{\partial F}{\partial c_i} \right)^2 (\sigma_i'^2 - \sigma_i^2) + \left(\frac{\partial F}{\partial c_i} \right)^2 (\delta_i'^2 - \delta_i^2) \right] \right\} \qquad (3\text{-}12)$$

For the above three equations [Equation (3-10)~(3-12)], σ_i' is the changed value that related to σ_i, while δ_i' is the changed value that related to δ_i. Their actual numerical value can be obtained with kinds of advanced measuring equipment in practical digital manufacturing environment. According to the calculated quality loss results in above three equations, an initial sorting work can be done. The quality loss ΔL_i can be regarded as the influence degree of a single alternative CF c_i affecting the whole CFs y_m at a higher assembly hierarchy. Then the top-ranked alternative CFs can be taken as the desired CF.

However, it is noted that only relying on the above identifying method, a missed or

Digital Assembly Coordination and Quality Controlling Technology
for Aeronautical Thin-walled Structures

multiple selections for CFs during the entire assembly process may happen. Due to the characteristics of the multi-hierarchy process for aircraft products and multi-station for assembly process, a situation frequently occurs, that is a component or a module would be taken as a special "part" in the assembly process at a higher assembly hierarchy. Then, the CFs at different assembly hierarchies or different manufacturing stages would interact and couple with each other, forming a complex evolution network, as shown in Figure 3-5.

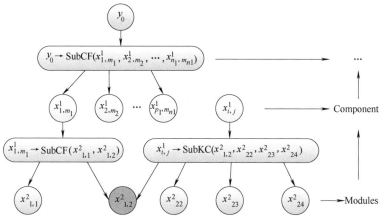

Figure 3-5   Evolution network for coordination features

In the evolution network for coordination features, the alternative CF $x_{1,2}^2$ belongs to a module assembly hierarchy, and its variation has influences on the quality loss of x_{1,m_1}^1 and $x_{i,j}^1$ that contained in the component assembly hierarchy. By applying qualitative analysis, $x_{1,2}^2$ is regarded as a key feature. However, to the mating features $x_{1,m_1}^1, x_{2,m_2}^1, \cdots, x_{n_1,m_{n1}}^1$ and $x_{i,j}^1$ that are located at different manufacturing stages, are contained in the component assembly hierarchy. These features work together to affect the variation of the CF y_0 at a higher assembly hierarchy. In the variation delivery process that aforementioned, due to a tight relationship with the CF y_0, the above mating features of different assembly hierarchies can be taken as the potential alternative CF set $\{c_1, c_2, \cdots, c_n\}$.

At the next stage, to identify the potential alternative CFs in the interaction and evolution network, the concept of affected degree is proposed. The quality loss ΔL_m that calculated in Equation (3-11) can be regarded as the sum of the influence degrees, which is caused by the entire relevant alternative CFs $c_1, c_2, \cdots, c_n$ at a lower assembly hierarchy. On the other hand, ΔL_m is also referred as the affected degree of the feature y_m at a higher assembly hierarchy. Then with the help of DEMATEL method that based on fuzzy theory, to accurately identify the CFs for aircraft product, the centrality degree index and the cause degree index are proposed for calculating the importance degree. Where the DEMATEL method is a systematic analysis and decision-making method, that is also an effective methodology to solve the complex group problems.

The centrality degree index, which is denoted as Z_i, ranges throughout the whole identification process. Where Z_i indicates the function of a certain potential alternative CF c_i in the overall assembly hierarchies. By summing the influence degree ΔL_i and the affected degree ΔL_m together, its numerical value can be calculated, as expressed in Equation (3-13).

$$Z_i = \Delta L_i + \Delta L_m \qquad (3\text{-}13)$$

The calculation result Z_i reflects the absolute importance of a certain feature $c_i (i = 1, 2, \cdots, n)$ in the alternative potential CF set $C = \{c_1, c_2, \cdots, c_n\}$. Contrasting to other potential alternative CFs at different assembly hierarchies, a relative larger calculation result indicates a stronger relationship of the mating feature c_i, and more attention should be paid on its precision in the manufacturing and assembly process. Conversely, a smaller value indicates a weaker relationship.

For the cause degree index, which is denoted as R_i, reflects the indirect importance of the alternative CF c_i in the potential feature set $c_1, c_2, \cdots, c_n$. R_i has a judging effect that ① whether a coordination feature influences other mating features at a higher assembly hierarchy more, or ② whether the coordination feature is easily affected by other mating features at a lower assembly hierarchy. It can be calculated by the difference between ΔL_i and ΔL_m, as shown in Equation (3-14).

$$R_i = \Delta L_i - \Delta L_m \qquad (3\text{-}14)$$

When the calculation result R_i is greater than zero, this situation indicates that c_i influences other features y_m more, making it an influencing factor. Conversely, if the calculation result R_i is smaller than zero, it means that c_i is an affected result relative to other mating features, making it an outcome-oriented factor. The result $R_i \prec 0$ also indicates that the potential alternative CFs at a lower assembly hierarchy should be controlled to meet the design and manufacturing requirements.

In summary, for the three adjacent assembly hierarchies that mentioned in Figure 3-3 and Figure 3-5, the comprehensive identification for CFs, based on a completed importance degree modeling, can be expressed in Equation (3-15) and (3-16).

$$\omega_i = Z_i (1 - R_i') / \sum_{i=1}^{n} [Z_i (1 - R_i')] \qquad (3\text{-}15)$$

$$R_i' = R_i \Big/ \left(\sum_{i=1}^{n} |R_i| \right) \qquad\qquad (3\text{-}16)$$

In the above two Equations, where ω_i represents the completed importance degree of a certain potential alternative CF c_i. The calculation results are the direct identification basis to determine whether a potential alternative CF is the desired CF or not.

However, with the Equation (3-3)~(3-16), only the identification process of the potential alternative CFs at a special assembly hierarchy, namely, the middle jth hierarchy, is completed by referencing the mating features at its two adjacent assembly hierarchies, i.e., i and k. But to the CFs at other assembly hierarchies, the identification process is the same as the jth assembly hierarchy, specifically, it involves calculating the complete importance degree to identify the CFs of products.

In order to gain a deeper understanding of the CFs identification process for an aircraft product, three explanations are provided here.

① Since the aircraft product CFs have obvious characteristics of multi-level and multi-station, the CFs that distributed on parts at the lowest assembly hierarchy cannot be decomposed downward anymore, resulting in the zero value of affected degree ΔL_m. So only the influence degree ΔL_i is taken into account when calculating their importance degrees. However, the product's top CFs are only affected by the CFs at the lower assembly hierarchies. Under this situation, the value of influence degree ΔL_i is zero. As a result, only the affected degree ΔL_m is taken into account when calculating their complete importance degrees.

② Because the CFs' variation have a direct impact on coordination accuracy, the geometric parameters should be preferential guaranteed to meet the design and assembly requirements. Therefore it is important to determine the number of CFs on parts or components in the practical digital manufacturing situations. An excessive number will increase the difficulty of manufacturing and assembly process. As a result, based on satisfying assembly coordination, for the parts with a simply structure, the less is the better; for the complex assemblies, a proper adequate identification decision can be made, but the number generally should not more than 5.

③ With the development of product's manufacturing process, the identified CFs may change at different assembly stages or stations, even for the same parts or components. To get accurate identification results, a repetitive calculating process should be done according to practical manufacturing situations.

3.6　Experimental Verification

In this section, a wing flap component would be taken as the object to verify feasibility of the proposed method. By analyzing the structure and assembly process of the experimental object, in the first engineering case, coordination features on parts that affecting the skin profile of a single component are identified. Then, another set of coordination features on different components of the whole wing flap component, which affect the flush coordination requirement between skins at a higher assembly hierarchy, are further identified. Finally some improved actions are taken to guarantee the coordination accuracy, and beneficial results are shown as a conclusion. To verify the methodology's feasibility, a type of wing flap component of certain aircraft is presented. The structure and assembly process of the experimental object can be shown in Section 2.4.

3.6.1　The First Study Case

To identify the CFs on parts that affect the skin profile, the first study case is taken for the inner wing flap component.

According to the function of wing flap component, taking its design structure (Figure 3-6) and assembly process into account, the top CFs of this component are determined with the qualitative method firstly. The results are skin profile, axis of hinge joint, coaxiality of hinge joints at different positions, and positions of the two ending ribs. Then, the top CFs are decomposed into parts, as shown in Figure 3-6. To be used in the further identification process, these mating features that distributed on each part can be taken as the potential alternative CF sets.

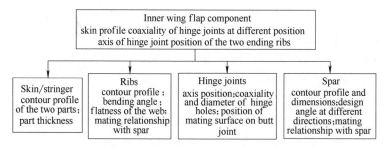

Figure 3-6　Decomposition of wing flap component

To get an accurate skin profile, structure of the related comprised parts, including spar, ribs, stringer, and skin panel, is analyzed. The detailed mating features F_{ij} of these parts that related to skin profile are shown in Figure 3-7. Where F_0 represents the datum feature on spar, i.e., the axis of hinge hole; F_{13} / F'_{13} and F_{15} / F'_{15} represent the inner and outer surfaces of spar, respectively; F'_{25} and F_{25} indicate the inner and outer flange surfaces of rib that have a mating relationship with spar; F'_{27} and F_{27} represent the inner and outer profiles of rib; F_{31} and F_{32} indicate the inner and outer profiles of stringer; F_{41} and F_{42} represent the inner and outer profiles of skin panel. The detailed constraint relationships, namely distance/angle/mating status among the mating features that depicted in Figure 3-7, are shown in Table 3-2.

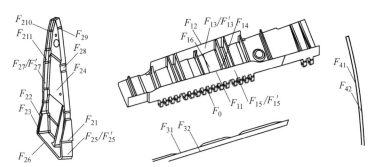

Figure 3-7　Mating features on different parts that related with skin profile

Table 3-2　Detailed constraint information among the comprised mating features

Mating feature		Relationship between features	Mating feature		Relationship between features
F_0	F_{11}	70mm	F_{25}	F'_{25}	3mm
F_0	F_{15}	0.05°	F_{21}	F_{26}	40mm
F_{11}	F_{12}	100mm	F_{22}	F_{16}	40mm
F_{13}	F'_{13}	2mm	F_{27}	F'_{27}	1.5mm
F_{15}	F'_{15}	2mm	F_{28}	F_{21}	190mm
F_{13}	F_{16}	133°	F_{29}	F_{21}	290mm
F_{15}	F_{16}	46°	F_{210}	F_{22}	220mm
F_{11}	F_{21}	0mm/0° (contact)	F_{211}	F_{22}	160mm
F_{12}	F_{22}	0mm/0° (contact)	F_{28}	F_{31}	0mm/0° (contact)
F_{13}	F_{23}	0mm/0° (contact)	F_{31}	F_{32}	1.5mm
F_{14}	F_{24}	0mm/0° (contact)	F_{29}	F_{41}	0mm/0° (contact)
F_{15}	F_{25}	0mm/0° (contact)	F_{32}	F_{41}	0mm/0° (contact)
F_{16}	F_{26}	0mm/0° (contact)			

Based on the geometric relationships among different mating features, four VPCs that have a direct relationship with the component's skin profile can be built, as shown in Figure 3-8. Where the first VPC represents the variation propagation process with the mating features that affect the skin of lower airfoil profile beyond the stringer region; the second VPC indicates the variation propagation process with the mating features that affect the skin of top airfoil profile beyond the stringer region; in the third VPC, the variation propagation process with the mating features that affect the skin of lower airfoil profile at the stringer region is expressed; the fourth VPC represents the variation propagation process with the mating features that affect the skin of top airfoil profile at the stringer region; and each $c_i(i=1,2,\cdots,23)$ indicates the constraint relationships among these mating features.

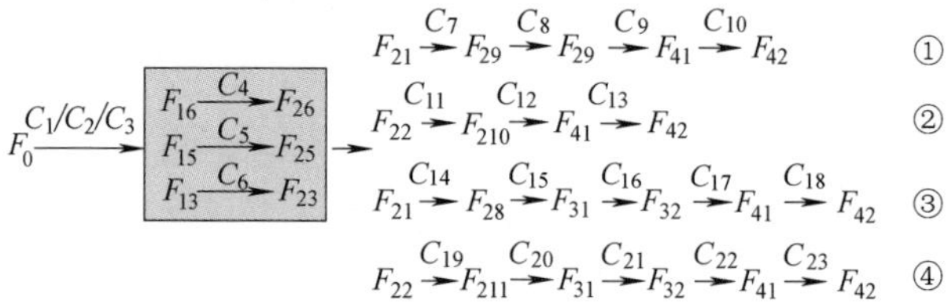

$$F_0 \xrightarrow{C_1/C_2/C_3} \begin{matrix} F_{16} \xrightarrow{C_4} F_{26} \\ F_{15} \xrightarrow{C_5} F_{25} \\ F_{13} \xrightarrow{C_6} F_{23} \end{matrix}$$

$$F_{21} \xrightarrow{C_7} F_{29} \xrightarrow{C_8} F_{29} \xrightarrow{C_9} F_{41} \xrightarrow{C_{10}} F_{42} \quad \textcircled{1}$$

$$F_{22} \xrightarrow{C_{11}} F_{210} \xrightarrow{C_{12}} F_{41} \xrightarrow{C_{13}} F_{42} \quad \textcircled{2}$$

$$F_{21} \xrightarrow{C_{14}} F_{28} \xrightarrow{C_{15}} F_{31} \xrightarrow{C_{16}} F_{32} \xrightarrow{C_{17}} F_{41} \xrightarrow{C_{18}} F_{42} \quad \textcircled{3}$$

$$F_{22} \xrightarrow{C_{19}} F_{211} \xrightarrow{C_{20}} F_{31} \xrightarrow{C_{21}} F_{32} \xrightarrow{C_{22}} F_{41} \xrightarrow{C_{23}} F_{42} \quad \textcircled{4}$$

Figure 3-8   Variation propagation chains relevant with the skin profile

It is mentioned that, for the complex comprehensive mating relationship between rib and spar, which contains three constraint relationships, i.e. C_4, C_5 and C_6. And a clearance error would occur at the mating area of F_{11}/F_{21} and F_{12}/F_{22}. By browsing the constraint relationships that contained in Table 3-2, the normal distances between F_{11} and F_{12}, F_{12} and F_{22} are the same, with a numerical value of $L_1=L_2=100\text{mm}$. Without considering the locating factors that contained in assembly process, to calculate the assembly coordination error between the rib and spar, a combination of thickness error and contour error are taken into account. The dimension error chain is given in Equation (3-17).

$$V_{\Sigma 1} = L_1 + \Delta_1 + \Delta_2 - L_2 - \Delta_1' - \Delta_2' \tag{3-17}$$

Where, Δ_1 and Δ_1' represent the contour error of mating features on the above two parts, each with a same variable range of $\pm0.01\text{mm}$. Δ_2 and Δ_2' represent the thickness error of these mating features respectively, with a range of $\pm0.004\text{mm}$. By applying the probabilistic method, the calculated result is shown in Equation (3-18).

$$(V_{\Sigma 1})_l^u = {}_{-0.006}^{+0.036}\text{mm} \tag{3-18}$$

Then, the angle deviation of the rib, which is caused by the mating error between rib and spar that perpendicular to the wing reference plane, can be calculated, as shown in Equation (3-19). Where, $L_{F_{21}-F_{26}}$ represents the length constraint between the mating areas of rib and spar, with a value of 40mm, as contained in Table 3-2.

$$\Delta_\theta = (V_{\Sigma 1})_i^u / (L_{F_{21}-F_{26}}) = [-0.00015°, +0.0009°] \qquad (3-19)$$

Based on the calculation results of $(V_{\Sigma 1})_i^u$ and Δ_θ, with the consideration of dimension constraint relationship between the features of F_{28} and F_{21}, the error at the feature F_{28}, which is perpendicular to the wing datum plane that caused by the angle deviation of the rib, can be calculated in Equation (3-20). It is noted that the feature F_{28} of the rib shares the same position as the stringer.

$$\Delta_{28} = L_{F_{21}-F_{28}} \times \Delta_{\theta 2} = [-0.029, +0.171]mm \qquad (3-20)$$

From the above analysis, we can know that the existence of form tolerances (contour error, thickness error, etc.) of the mating features would influence the final variation propagation results of the CFs. By taking the form tolerance and dimension tolerance into overall consideration, with the using of Equation (3-3)~(3-12), quality loss of the component's skin profile that caused by the precision variation in relative alternative CFs is calculated. The calculation parameters and ranking results for the third VPC are shown as an example in Table 3-3.

Table 3-3   Calculation parameters and ranking result for the third VPC

Constraint relationship	Target value/mm		Practical measured value/mm		Variation propagation coefficient	Quality loss	Ranking result
	σ_i	δ_i	σ_i'	δ_i'		ΔL_m	
C_{14}	0.01	0	0.03	0.005	0.05	0.86	5
C_{15}	0.004+0.008	0	0.015	0.007	1	1.89	4
C_{16}	0.004+0.008	0	0.015	0.008	1	2.12	3
C_{17}	0.008	0	0.010	0.006	1	2.37	1
C_{18}	0.004+0.008	0	0.015	0.008	1	2.25	2

With the analysis work on the quality loss calculation results, a ranking work is done. It is shown that the constraint relationships of C_{17}, C_{18}, and C_{16} have a greater influence on skin profile compared to the other two relationships of C_{14} and C_{15}. The three constraint relationships are the mating error between skin and stringer, the thickness error and contour error of the skin part, and the thickness error and contour error of the stringer part. Thus, a conclusion is drawn that the mating features related to the constraint relationships of C_{17}, C_{18}, and C_{16} are taken as CFs, and much attention should be paid in their manufacturing process to meet the precision requirements.

On the other hand, by reviewing the official documents that used in assembly process, we find that the assembly precision requirement of the skin profile is within the range of [−0.60, +0.40]mm. But for the calculation results in Equation (3-20), we also find that the comprised error item Δ_{28} of the rib profile's variation occupies a large portion of the required assembly accuracy. Obviously, the mating features that have a relationship with Δ_{28} are taken as the CFs, i.e., the outer surface at the bottom area of the rib and the corresponding inner surface of the spar. Their dimensions and shapes should also be carefully controlled in manufacturing process.

To verify the methodology's feasibility, in the practical assembly process, it is discovered that four main sources are related to the incoordination problems in this study case: the inapposite tolerance requirements between rib and spar; an overage of manufacturing tolerance on the flange edge of the ribs; an excessive tolerance at the mating regions on different parts; and a poor installation precision of the assembly jig. Without taking the locating and jointing factors during the assembly process into account, the CFs identified here is in accordance with the practical assembly situation, verifying the effectiveness of the methodology that proposed in this chapter successfully.

3.6.2 The Second Study Case

In the first case mentioned above, the identification process of a single inner wing flap component is presented. But to the whole set of wing flap components, which are located at a higher assembly hierarchy and another assembly stage with different assembly jigs, the flush coordination accuracy between skins of inner component and outer component is also required to guarantee in the assembly document, as shown in Figure 3-9. Where, V_{flush} is the offset or altitude difference between the two different surfaces. The grid portion that marked in the figure is the coordination region, and the direction of flush is perpendicular to the skin surface of the ending ribs in this case.

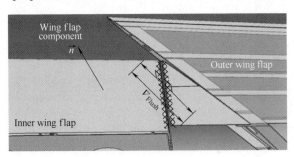

Figure 3-9 Flush requirement between different components (See the color illustration)

Digital Assembly Coordination and Quality Controlling Technology
for Aeronautical Thin-walled Structures

To ensure the top CF of the skin profile in the complete wing flap component, with the qualitative analysis, skin profile and positions precision for the ending ribs of the comprised inner and outer components are taken as the CFs in each component. And the decomposition of the flush coordination requirement at a higher assembly hierarchy can be shown in Figure 3-10.

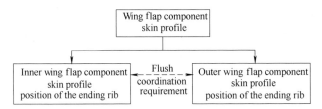

Figure 3-10   Decomposition of the flush coordination requirement between different components

In this section, to identify the CFs that having a relationship with flush coordination requirement, the quality loss ΔL_m of skin profile, caused by the variation of the potential alternative CFs, should be summed together, as previously mentioned in Table 3-3. Where, $m = 1, 2, \cdots, 5$. And the total influence degree for comprised parts of the inner component is calculated as $\Delta L_1 = 6.57$. Following the same procedure, the outer wing flap component is analyzed. By calculating the quality loss $\Delta L'_m$ of the skin profile in the practical manufacturing environment, the total influence degree of the comprised parts of the outer component has a value of 5.83mm, which is denoted as ΔL_2. And the influence degree of the inner component that affecting the whole wing flap component is calculated as $\Delta L_3 = 8.40$, and the influence degree of the outer component that affecting the whole wing flap component is calculated as $\Delta L_4 = 9.32$. Based on the calculation results of $\Delta L_i (i = 1, 2, \cdots, 4)$, according to the calculation method of centrality degree index Z_i and cause degree index R_i as mentioned in Equation (3-13) and (3-14), centrality degree index Z_i of the flush between two skin profiles that caused by the inner component is calculated as $Z_1 = \Delta L_1 + \Delta L_3 = 14.97$; Z_i that caused by the outer component is calculated as $Z_2 = \Delta L_2 + \Delta L_4 = 15.15$; cause degree index R_i of the flush between two skin profiles that caused by the inner component is calculated as $R_1 = \Delta L_1 - \Delta L_3 = -1.83$; R_i that caused by the outer component is calculated as $R_2 = \Delta L_2 - \Delta L_4 = -3.49$. Then with the help of Equation (3-15) and (3-16), the results of complete importance degree of the skin profiles for the inner components and the outer component are $\omega_1 = 0.45$, and $\omega_2 = 0.55$, respectively.

According to the analysis of calculation results of Z_1, Z_2, R_1, R_2, ω_1 and ω_2

in this case, we find that the values of centrality degree indexes Z_1 and Z_2 are almost the same, and the difference between their complete importance degree ω_1 and ω_2 is very small. The above conclusion reflects the accuracy of the qualitative identification results, i.e. the skin profiles of the comprised inner and outer components are taken as the CFs that needed in each component. Meanwhile, the cause degree indexes R_1 and R_2 of the two comprised components are both negative, indicating that they are both the outcome-oriented factors. This situation also means that the skin profiles of the two components are deeply affected by their respective comprised parts. Consequently, a conclusion is drawn that the flush of the whole wing component has a direct relationship with the skin profiles of the ending ribs on each single comprised component, which should be paid much attention for controlling their assembly process.

In practical assembly process of the ending ribs on the above four comprised components (Figure 3-11), the measuring precision information of the components is shown in Table 3-4. Where the measuring data of five wing flap components' skin profiles with laser tracker (API track III) is contained; a, b, c, and d represent the four ending ribs on the four single components that to be assembled for a whole wing flap component, which have a

Figure 3-11　Practical flexible assembly site

relationship with the flush coordination accuracy. A total 20 sets of precision data is to be analyzed in Table 3-4.

Table 3-4　Measuring profile precision for ending ribs

Sample		Precision/mm	Sample		Precision/mm
The 1st wing flap component	a	0.36	The 6th wing flap component	a	0.18
	b	0.38		b	0.24
	c	0.40		c	0.29
	d	0.08		d	0.17
The 3rd wing flap component	a	0.27	The 8th wing flap component	a	0.28
	b	0.35		b	0.26
	c	0.32		c	0.35
	d	0.33		d	0.20
The 4th wing flap component	a	0.35			
	b	0.39			
	c	0.29			
	d	0.31			

By combining the 20 precision data of the four ending ribs on five wing flap components, their variation situation based on statistical process control (SPC) method[27], can be shown in Figure 3-12.

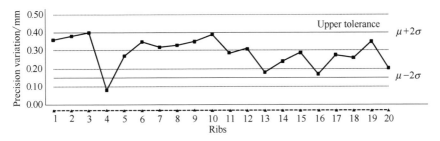

Figure 3-12   Precision variation at the 20 ending ribs' profile

By browsing the official documents that used in assembly process again, we find that the required assembly precision that has a range of [−0.60, +0.40]mm. According to the practical precision variation, we can know that the assembly precision data for ending ribs is not located at the center region (−0.1mm) of the assembly precision range, but at the top area of the tolerance zone. And we can also find that these data have a bigger moving range relative to each other, as shown in Figure 3-13 (colored in purple), reflecting an unstable assembly process.

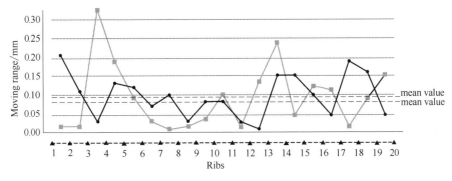

Figure 3-13   Moving range of the precision data (See the color illustration)

To stabilize the assembly process and bring the precision data closer to center line of the required assembly precision range, the reasons of the above phenomena observed in Figure 3-13 and Figure 3-14 are analyzed. In practical assembly site, we find that the locations of ending ribs rely on two moveable stop blocks, which is attached to the contour board, and the locating process is a manual adjusting work. Based on this situation, an improved work is done. Solutions involves a special device is designed and manufactured to locate the ending rib stably and accurately. When the new assembly work is completed, to collect the assembly precision data of the skin profile at the positions of different ending ribs for another five wing flap components, another measuring work is done, as shown in Table 3-5. The precision variation situation of these data can be shown in Figure 3-14, with the moving range (colored in blue) is shown in Figure 3-13.

Table 3-5　Measuring precision after adding the special locater for ending ribs

Sample		Precision/mm	Sample		Precision/mm
The 9th wing flap component	a	−0.35	The 12th wing flap component	a	−0.19
	b	−0.14		b	−0.34
	c	−0.25		c	−0.19
	d	−0.22		d	−0.29
The 10th wing flap component	a	−0.35	The 13th wing flap component	a	−0.34
	b	−0.27		b	−0.15
	c	−0.34		c	−0.31
	d	−0.24		d	−0.36
The 11th wing flap component	a	−0.21			
	b	−0.29			
	c	−0.21			
	d	−0.18			

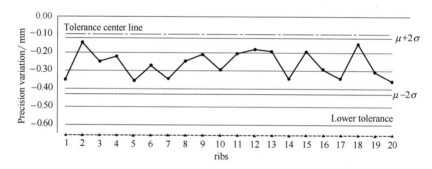

Figure 3-14　Precision variation at another 20 ending ribs' profile

By summarizing the practical measuring data mentioned in Figure 3-12~3-14, it can be known that after adding the special ending rib locators, ① the deviation of the skin profile at the ending ribs is more accurate than before, and ② the tolerance is also closer to center line of the required assembly precision range, as shown in Figure 3-14. The situation (a relatively accurate skin profile and a relatively smaller skin profile compared to the ideal profile) would bring an enhancement on aircraft's flight performance. Furthermore, by calculating the mean value of the moving range for each precision date that shown in Figure 3-13, we find that they are almost the same with each other, as shown by the black line and gray line. However, after calculating the standard deviation of the moving range for each precision date shown in Figure 3-13, we also find that the standard deviation is 0.078 for the first five wing flap components, and the standard deviation is 0.057 for the last five components. By comparing these two results, we can know that there is an enhancement of 26.9% in the stability during assembly process, which would lead to a higher assembly quality.

3.7 Summary

① Based on the multi-hierarchy and multi-station assembly process for aircraft products, to identify the CFs that have a relationship with coordination accuracy requirements, a comprehensive method based on complete importance modeling is developed in digital manufacturing environment. Firstly, the interaction and evaluation relationships between assemblies at different manufacturing stages are analyzed. Secondly, the quality loss at three adjacent assembly hierarchies is calculated. Thirdly, the centrality degree index and cause degree index are proposed for calculating the complete importance degree, and then coordination features among different assembly hierarchies are identified eventually.

② By using the error correction coefficient H, the reliability of calculation results can be enhanced, making the identified results more accordant with practical assembly circumstances.

③ Certain type of wing flap components are presented to indicate the method-logy's feasibility, and coordination features at different assembly hierarchies that affect the skin profile of a single component are successfully identified. Furthermore, coor-dination features, which have a relationship with the flush coordination accuracy between the skins profile of the whole wing component, are also identified. According to the precision data with a statistical analysis, first, the identified CFs are in accordance with the practical assembly application situations; second, some improved actions are taken to guarantee the coordination accuracy. Benefitical results showed that the deviation of skin profile is smaller than before, and the tolerance is closer to center line of the required assembly precision range. Furthermore, the stability in assembly process is increased by 26.9%, which would bring a higher assembly quality and an enhancement in aircraft's flight performance.

References

[1] Bullen G. Automated/Mechanized Drilling and Countersinking of Airframes[M]. Warrendale: SAE International Press, 2013.

[2] Muelaner J, Kayani A, Martin O, et al. Measurement assisted assembly and the roadmap to part-to-part assembly[C]. 7th International Conference on Digital Enterprise Technology, University of Bath, 2011:11-19.

[3] Jin Y, Abella R, Ares E, et al. Modeling and digital tool development of a new similarity metric for aerospace production[J]. International Journal of Advanced Manufacturing Technology, 2013, 69(1-4): 777-795.

[4] Quintana V, Rivest L, Pellerin R, et al. Will Model-based Definition replace engineering drawings throughout the product lifecycle? A global perspective from aerospace industry[J]. Computers in Industry, 2010, 61(5): 497-508.

[5] Irving L, Ratchev S, Popov A, et al. Implementing determinate assembly for the leading edge sub-assembly of aircraft wing manufacture[J]. SAE International Journal of Aerospace, 2014, 7(2): 246-254.

[6] Yang Z, Mcwilliam S, Popov A, et al. A probabilistic approach to variation propagation control for straight build in mechanical assembly[J]. International Journal of Advanced Manufacturing Technology, 2013, 64(5-8): 1029-1047

[7] Liu, X, Ni, Z, Liu J, et al. Assembly process modeling mechanism based on the product hierarchy[J]. International Journal of Advanced Manufacturing Technology, 2016, 82(1-4):1-15.

[8] Thornton A, Donnelly S, Ertan B. More than just robust design: Why product development organizations still contend with variation and its impact on quality[J]. Research in Engineering Design, 2000, 12(3): 127-143.

[9] Whitney D. The role of key characteristics in the design of mechanical assemblies[J]. Assembly Automation, 2006, 26(4): 315-322.

[10] Ola A. Quality modeling case study at GKN Aerospace Sweden[D]. Gothenburg: Chalmers University of Technology, 2015.

[11] Mathieu L, Marguet B. Integrated design method to improve producibility based on product key characteristics and assembly sequences[J]. CIRP Annals-Manufacturing Technology, 2001, 50(1): 85-88.

[12] Andolfatto L, Thiébaut F, Lartigue C, et al. Quality- and cost-driven assembly technique selection and geometrical tolerance allocation for mechanical structure assembly[J]. Journal of Manufacturing Systems, 2014, 33(1): 103-115.

[13] Johansson P, Bergman B, Svensson T, et al. A robustness approach to reliability[J]. Quality & Reliability Engineering International, 2013, 29(1): 17-32.

[14] Tang W, Li Y, Zhang J, et al. A systematic top-down approach for the identification and decomposition of product key characteristics[J]. Proceedings of the Institution of Mechanical Engineers, Part B: Journal of Engineering Manufacture, 2016, 228(10): 1305-1313.

[15] French D, Wu Y, Li Y. Identifying the relative importance of stock characteristics[J]. Journal of Multinational Financial Management, 2016, 34(3): 80-91.

[16] Elfenbein H, Eisenkraft N, Curhan J, et al. On the relative importance of individual-level characteristics and dyadic interaction effects in negotiations: Variance partitioning evidence from a twins study[J]. The Journal of Applied Psychology, 2018, 103(1): 88-96.

[17] Campos A, Johnson R, Kennedy J. System and method for wiring an aircraft[P]. US2016004863, USA. 2017-08-08.

[18] Eric P, Christophe M, Denis G. Key tool and composite panel, in particular for an aircraft engine nacelle, produced using such a tool FR3009992A1[P]. 2015-03-06.

[19] Guo F, Wang Z, Kang Y, et al. Positioning method and assembly precision for aircraft wing skin[J]. Proceedings of the Institution of Mechanical Engineers, Part B. Journal of engineering manufacture, 2018, 232(2): 317-327.

[20] Jareteg C, Wärmefjord K, Söderberg R, et al. Variation simulation for composite parts and assemblies including variation in fiber orientation and thickness[J]. Procedia CIRP, 2014, 23(12): 235-240.

[21] Karlsson J, Markmann C, Alam M, et al. Parameter influence on the laser weld geometry documented by the matrix flow chart[J]. Physics Procedia, 2010, 5(1): 183-192.

[22] Madrid J, Soderberg R, Vallhagen J, et al. Development of a conceptual framework to assess producibility for fabricated aerospace components[J]. Procedia Cirp, 2016, 41: 681-686.

[23] Ma L, Mao J, Fan H. Key Quality Characteristics identification method for mechanical product[J]. Procedia Cirp, 2016, 56: 50-54.

[24] Han X, Li R, Wang J, et al. Identification of key design characteristics for complex product adaptive design[J]. International Journal of Advanced Manufacturing Technology, 2017, 95(1): 1215-1231.

[25] Falatoonitoosi E, Leman Z, And S, et al. Decision-making trial and evaluation laboratory[J]. Research Journal of Applied Sciences Engineering & Technology, 2013, 5(13): 3476-3480.

[26] Narasimhan K. Quality From Customer Needs to Customer Satisfaction (3rd Edition)[J]. The TQM Journal, 2011, 23(3): 358-359.

[27] Fred W. Statistical Process Control in Manufacturing Practice[M]. Boca Raton: CRCPress, 1997.

Chapter 4
Positioning Error Guarantee Modeling for Flexible Assembly Tooling

4.1 Introduction and Related Work

4.1.1 Background

Aircraft have strict requirements of aerodynamic profile accuracy. However, its constituent parts, such as the skin, frame, and stringer, having a weak rigidity[1,2]. Differing from common mechanical products, an aircraft needs to keep the designed structure shape/size, while meeting the requirements of ① design accuracy[3], and ② interchangeable consistency between different assemblies[4]. Due to the characteristic of prone deformation, its assembly work needs the assistance of a large number of special locating devices[5,6]. And assembly tooling is used to support components while they are being processed, and to locate/joint different parts together in correct relative positions under the design coordinate system[7,8]. For the traditional assembly method, the assembly work is carried out on the rigid jig, with the dimension/shape/position transfer basis of standard hard master tooling[9]. However, the assembly frame is specially manufactured, and is not suitable for assembling other different aircraft components[10]. As the assembly precision and the consistency between different assemblies become more important, correspondingly, there also has a positioning error enhancement for assembly tooling.

Considering the requirements of precise assembly for new generation aircraft, the development of flexible assembly tooling technology can realize this goal. The tooling's configuration can be adjusted automatically to fit the changes of different assembly environments, and assist in the assembly of products with different shapes, sizes, and even structures. With its high locating accuracy, the accurate dimensional size/shape/position and good assembly precision could be gained. In practical assembly process, the ability of accurate locating for tooling is an important determinant factor for high precise

Digital Assembly Coordination and Quality Controlling Technology
for Aeronautical Thin-walled Structures

assembly. How to guarantee the positioning error of aircraft flexible assembly tooling with a precise and effective mode, is the main purpose of this chapter.

4.1.2 Application and Structure of Flexible Assembly Tooling

For aircraft products, flexible assembly tooling has gained more attention and has been applied in actual production. In order to assembly different components, Li[11] developed a digital flexible pre-assembly tooling system for locating and clamping different fuselage panels. Considering multiple constraints, Tadic[12] proposed a general model for assembling workpieces with complex geometry by two skewed holes. Erdem[13] and Keller[14] proposed an automated flexible assembly tooling for wing box assembly with force-controlled adjustments, and took the tooling with hexapod solutions as an enabler in the assembly process of cross section structures. In the case study, Zheng[15] presented a rapid configuration for box-joint assembly jigs, which was used in the airbus A-380 subassembly system. In addition, the corresponding methodology for high accuracy installation of sustainable jigs and fixtures was also proposed. Moreover, Olabanji[16] provided relevant information on the design analysis of a novel reconfigurable assembly fixture for press brakes, and then conducted a detailed mechanical design method for the critical parts of the reconfigurable assembly fixture.

For the assembly work of aircraft components, the flexible assembly tooling system normally comprises a lot of locating units. In this chapter, in order to realize the function of reconfigurable positioning, one locating unit with three motion layers is taken as the research object. The unit can accomplish the movement along three perpendicular directions, i.e., X, Y, and Z. Each motion direction is constituted of a servo drive system, high-precision guiding system with certain preloading force, ball screw drive system, and other accessories, as shown in Figure 4-1.

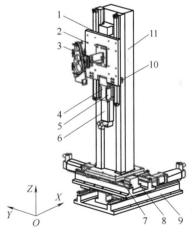

Figure 4-1 Structure of flexible locating unit

1—guiding rail; 2—connecting plate; 3—end locating effector; 4—connecting flange; 5—shaft coupling; 6—servo motor; 7—Ball screw; 8—Bearing seat; 9—column base; 10—Pneumatic clamping parts; 11—column body

The working characteristics of the above locating unit can be described as follows. In the flexible locating process, according to motion data provided by DELMIA simulation, the locating motion starts from the defined origin point along each direction within the tooling coordinate system. Following a given motion sequence and NC program, the end locating effectors are driven to move at a low speed (defined as 0.02m/s) to any spatial locating position within the effective stroke range. With the assistance of laser tracker, the position of key reference OTP (optical tooling point) for the entire locating unit can be recorded in real-time. Then, the actual spatial position deviation can be compensated with the difference value between measured results and theoretical position. Based on the precise adjustment on the positioning error along three rectangular coordinate directions, the locating and clamping work for aircraft parts can be carried out, when the unit's variation remains within an allowable range of 0.05mm.

4.1.3　Positioning Error Guarantee Method with Error Transfer Modeling

By summarizing the above structures and working characteristics, it is found that, because of the complicated structure and the diverse geometric forms of the aircraft products, the corresponding assembly tooling system usually has a complex structure. Taking the factors of manufacturing and installation error items for the locating unit into account, it is inevitable that the tooling will generate different positioning deviations based on different flexible locating positions and postures. For the technology of guaranteeing tooling positioning accuracy, in order to minimize the tolerance budget which is generated by the positioning variations of jigs and fixtures, Jamshidi[17] presented that, it was necessary to improve the installation accuracy of the jigs and fixtures to a high level, especially in aerospace industry. For the multiple influence factors on positioning error, Ramesh[18] stated that the geometric errors account for the largest proportion. Considering the locating function with three motion axes, Tian[19] proposed a general error modeling method that includes 21 geometric error terms. With error-differential relation theory and the concept of error weight coefficient, Lin[20] proposed an error compensation mathematical model that reveals the changing principle between the locating error of positioning platform and the geometric parameters. Due to many error sources for aircraft panel assembly process, by referring the concept of modularization, Zhao[21] proposed a comprehensive kinematic model for the dual-machine system, which contains the direct calibration based on laser tracker and error compensation method based on a special designed instrument. Deng[22] developed a linear stage by improving its structure to reduce the error sources, and proposed a macro-micro compensation method for the straightness motion error and positioning error. However, the above error modeling methods separated geometric error from deformation error, and Patel[23] presented that they cannot reflect the error transfer principle for the locating mechanism accurately. In order to calculate the deformation error caused by external loads, Grandjean[24] proposed an original procedure to systematically analyze and quantify the assembly of parts with form/position defects and deformable contacting surfaces. With the idea that there is a coupling relationship between geometric error and deformation error, Schleich[25] proposed an method to determine the impact of geometric deviations on the structural performance, in which the elastic deformation was expressed in tolerance domain by conventional tolerance expression. Then, in order to resolve this complex coupling relation, Lustig[26] presented a detailed procedure that containing five steps: discrete processing of contact surface, determining the type of surface contour function, constructing the surface contour function, function superposition for surface deviation and elastic deformation, and calculating the statistical moment. However, with the comprehensive solution of geometric modeling and deformation analysis, the realization of precise positioning for the locating unit still cannot be accomplished, due to the simplification of the error transfer modeling process, and the complex relationships of various time-varying parameters[27] in practical assembly site.

4.1.4 Positioning Error Guarantee Method Considering Practical Assembly Conditions

In practical assembly process, there are many time-varying influence factors of positioning error, such as assembly forces, temperature changes, different locating spatial positions, different motion distances, the changes of assembly objects, and the dynamic loads caused by the increasing product weight. As a result, the above solution based on geometric error transfer mechanism modeling cannot fit the practical assembly conditions. In order to analyze the strong nonlinear relationships of the comprised error items for the locating parts and motion modules, a locating motion model with sufficient learning ability for guaranteeing positioning error is required. With the measurement error data[28] obtaining from the practical assembly site, Wang[29] proposed an error separation algorithm based on the principle of sequential multi-alteration measurement, in which the laser tracker was used to measure the same motion trajectory of the linear axis and rotary axis at different base stations. Nubiola[30] presented an error model that takes 25 geometric error parameters and 4 error parameters into account, adopting the least squares optimization method, with the purpose of finding the 29 error parameters that best fit the error data acquired with a laser tracker. For the sake of identifying error parameters of the rotational axis along the error sensitive directions, He[31] proposed a simplified model accounts for spatial error, which contains the differential operator and a Jacobian matrix of differential motion along each motion axis. Due to the existence of a large number of time-varying parameters, it is always difficult to establish and solve the motion model. In order to minimize the positioning error of a high precision slide in the CNC turning machine, Vinod[32] demonstrated an adaptive and active compensation method for positioning error, and developed a module for real time computation and compensation with back propagation feed-forward neural network. With the means of data processing, Nguyen[33] proposed a calibration method for enhancing robot position accuracy, in which the extended Kalman filtering (EKF) algorithm was adopted for identifying geometric parameters, and the artificial neural network (ANN) method was applied to compensate the un-modeled errors.

The above literatures have conducted an in-depth and systematic analysis of error measurement, identification, separation, modeling mechanism and empirical analysis, as well as the coupling relationships between different error items, providing beneficial research ideas for guaranteeing positioning accuracy of aircraft flexible assembly tooling. However, because of the existence of nonlinear and coupling relationships, the above characteristics are not the same for different locating units. As a result, a locating motion model with sufficient learning ability, excellent prediction ability, and good interpretability effects is needed. Due to lack of the above comprehensive model, a real-time assistance from the laser tracker is often needed for guaranteeing positioning accuracy. Although the laser tracker is taken as a necessary auxiliary tool in practical assembly sites, it would cause low assembly efficiency and a poor application effect.

4.1.5　Structure of the Chapter

In this chapter, a method with two-stage compensation strategy for aircraft flexible assembly tooling is proposed, in which the actual geometric status and the time-varying state parameters for practical assembly conditions are considered, with the purpose of guaranteeing the positioning error. The remaining sections can be described in Figure 4-2, which shows the specific research route. Based on the practical measurement require-ments, Section 4.2 gives the establishment of assembly measurement field, Section 4.3 shows the accurate measurement of the actual error status and assembly conditions. Considering the actual error status and deformation analysis, Section 4.4 presents the error transfer mechanism modeling at pre-compensation stage. In Section 4.5, considering the time-varying sample data from practical assembly site, the ELM parameter identification method for SLFN motion prediction model is presented. Combining the advantages of mechanism model and error measurement data model, Section 4.6 presents the accurate compensation stage. To verify the methodology's feasibility, three progressive experiments are designed in Sections 4.4~4.6, respectively.

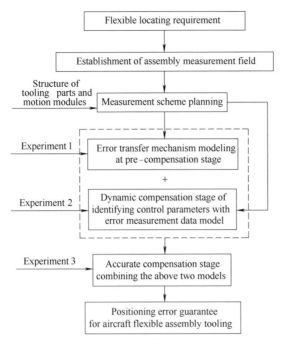

Figure 4-2   Research route for positioning error guaranteeing with
two-stage compensation strategy

4.2   Establishment of Assembly Measurement Field

Flexible assembly tooling is a motion mechanism comprises many connecting parts and linear motion modules. Its positioning accuracy is affected by the actual status of structural components and various time-varying parameters. To ensure the positioning error of the tooling, it is necessary to establish the assembly measuring field firstly[34]. The field is considered as the key link to realize the transmission of the product design data to the practical objects in digital form. Then, the fast and accurate measurements of geometric states and geometric relationships can be realized. To be more specific, the measuring field has three main functions to satisfy the procedure of flexible locating and assembly. The first function is the accurate collection of the actual mating precision conditions for the comprised parts and motion modules, and the working conditions during flexible assembly processes (such as drilling force, riveting force, and the forces applied on the positioning/clamping device caused by the increasing products weight). With the above actions, physical deformation error of the flexible tooling can be taken as a source of assembly deviation model. It would be much helpful for the pre-compensation stage that considers actual error status and mechanical analysis. Considering the flexible positioning requirements for multi components, the second function is to acquire the time-varying status parameters that affect the tooling's positioning accuracy in the flexible locating process. More specifically, the dynamic compensation stage based on real-time measurement can be carried out, with the real-time identification of motion control parameters for each axis of the locating unit. In addition, during the practical assembly process, the actual manufacturing accuracy and assembly status of the product can be measured in real-time, i.e., the third function of the measuring field. With the above operations, positioning error of the tooling can be reconstructed to match the actual product precision state. Then the ultimate goal of precise assembly could be achieved by adjusting the variation status of the flexible tooling.

In conclusion, the main purpose or requirements of constructing an assembly measurement field is to realize the perception and measurement of different time-varying state parameters, such as tooling status, product state, environmental state, and assembly stress/deformation. However, the construction scheme of measuring field directly affects its working mode and precision, showing a huge influence on assembly efficiency and assembly quality. Measurement process planning mainly includes the measurement equipment layout, measurement sequence planning, measurement requirements formulation, and other aspects. In practical assembly site, for the locating units with three motion axes, the sensors or measuring equipment used for establishing assembly measuring field, are shown in Figure 4-3.

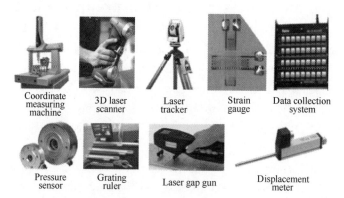

Figure 4-3   The sensor or measuring equipment used for establishing assembly measuring field

4.3   Accurate Measurement of Actual Error Status and Assembly Conditions

In the measuring field, with the combination use of the above equipment, the perception and measurement operations in the flexible assembly and manufacturing process can be realized. To be more specific, the accurate measurement scheme for various equipment is analyzed as follows.

① The geometric precision of the locating unit, mainly includes the precision of moving parts/modules and fixed parts. For moving parts, the axial feed error and installation posture error for the ball screw drive system in each moving direction, and the installation orientation error for the guiding rail drive system are included. For fixed parts, the corresponding error items mainly include the manufacturing/installation accuracy of the connecting plate between each motion direction, and the error for end locating effectors, i.e., the locators. In order to obtain the actual mating status of the tooling's comprised structures and the drive system, one specific method, the fitting and deviation feature extraction steps that related to the measured point cloud, is adopted to modify the shape tolerance complying with the practical situation. The above operation can be accomplished with the help of coordinate measuring machine, 3D laser scanner, and other tools. The detailed data processing and feature extraction procedures are shown in Figure 4-4.

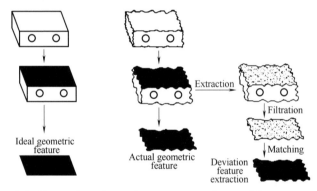

Figure 4-4   Processing of actual measurement data and feature extraction

② For the practical assembly conditions in the flexible locating process, key factors include the assembly force, the dynamic load, the changes of end locating effectors as assembling different components, and the deformation error, should be the major concerns. In order to measure the above geometric and physical quantities, the installation positions of different measuring devices are located near the end locating effectors. The pressure sensor and strain gauge are taken as the measuring basis for obtaining the assembly force and physical deformation in real-time. By measuring the key points that distributed on the end locating effectors, the actual assembly position of the tooling can be gained with the help of laser tracker. The above measuring scene for the locating unit can be shown in Figure 4-5. Based on the practical assembly data mentioned above, compensation can be applied to the positioning error, which is calculated by theory analysis with error mechanism modeling. It is mentioned that when using the laser tracker to measure the common coordinate datum points, at least three numbers of them are required.

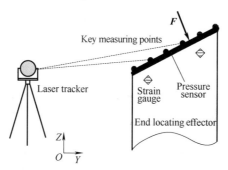

Figure 4-5   Measurement on the practical assembly conditions

③ In the measurement field, with the purpose of realizing the acquisition and measurement of time-varying status parameters[35] that related to flexible locating unit, specific displacement sensors, such as grating ruler, force sensor, and displacement meter (Figure 4-3), are used to obtain the practical working state of the locators. In addition, the temperature/humidity sensor is used to obtain the ambient assembly environment. The relevant data types for the locating unit, are mainly shown in Table 4-1.

Table 4-1   The type of sensing data associated with the locating unit

Attribute	Meaning	Notes
ITEM1	running status	0: unknown state; 1: to run; 2: in service; 3: fault; 4: emergency stop
ITEM2	Position in x-axis	Unit: mm
ITEM3	Position in y-axis	Unit: mm
ITEM4	Position in z-axis	Unit: mm
ITEM5	Bearing force in x-axis	Unit: N
ITEM 6	Bearing force in y-axis	Unit: N
ITEM 7	Bearing force in z-axis	Unit: N
ITEM 8	Whether the locating end is clamped	0: unlocked state; 1: locked state
ITEM 9	Power feedback signal for the control cabinet	Bool value
ITEM 10	Lock feedback signal in x-axis	Bool value
ITEM 11	Lock feedback signal in y-axis	Bool value
ITEM 12	Lock feedback signal in z-axis	Bool value
ITEM 13	Motor state feedback signal in x-axis	Bool value
ITEM 14	Motor state feedback signal in y-axis	Bool value
ITEM 15	Motor state feedback signal in z-axis	Bool value
ITEM 16	Control card status	Bool value
ITEM 17	Control network status	Bool value

Digital Assembly Coordination and Quality Controlling Technology for Aeronautical Thin-walled Structures

④ In the flexible locating process, for the actual manufacturing accuracy and assembly error status of the product, the measurement equipment, such as 3D scanner and laser tracker, can be used to obtain the physical profile deviation data of the curved surfaces and hinge joints. The above operations are taken as solutions of surface fitting or curve fitting based on the distributed measurement points, as shown in Figure 4-6.

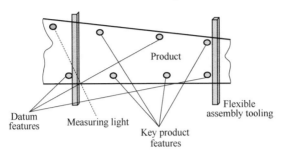

Figure 4-6   Measurement of the product

4.4   Pre-compensation Stage Considering Actual Error Status and Mechanical Analysis

In the process of flexible assembly, the ability of precise positioning of tooling is an important and determinant factor related to product assembly accuracy. By analyzing the factors causing the positioning error of the locating unit, based on the idea of geometric error modeling and finite element analysis[36], positioning accuracy for aircraft flexible assembly tooling with three-axis is guaranteed at this pre-compensation stage. Based on theoretical analysis, the purpose of this stage is to reveal the error transfer and accumulation mechanism of the comprised parts and motion modules. For this stage, the pre-compensation model would lay a foundation for the precise and effective compensation of positioning error in the next accurate compensation stage, where various time varying parameters and actual assembly conditions are concerned in Section 4.5.

4.4.1   Positioning Error Analysis

According to the structure characteristics of aircraft flexible tooling and its assembly/positioning functions, it can be found that there are no external loads/forces during the installation process for the positioning unit. However, the following factors, such as operating conditions, manufacturing/installation/geometric accuracy of the tooling's comprised parts and modules, and control accuracy of the locating unit, they all have an impact on the positioning error. According to statistical analysis, the principal factor that affecting the final positioning error is the geometric error of the comprised motion parts/modules. To be more specific, it mainly refers to the axial positioning error of ball screw drive system and its lead error, the installation orientation error of guiding drive system, and errors in other connecting parts, such as the connecting plate and jointing flange. It is mentioned that not all of the practical factors can be taken into account in the mechanism modeling, for example, the time-varying parameters. However, in order to facilitate more accurate error transfer mechanism modeling for positioning error, the following five assumptions are made.

① During the working process of flexible locating unit, positioning error of the end locating effectors, i.e., ΔP that caused by the displacement and orientation errors along the X, Y and Z motion directions, are independent of each other. And the three motion directions are controlled separately.

② In order to measure the position of locating unit, two or more OTP points are usually placed on the end locating effectors. Because there is a separation surface that connecting the end effector to its support bracket, the installation accuracy can be easily guaranteed with the help of laser tracker and necessary manual adjustments. However, considering the locating unit is a rigid body and the end effector is firmly connected to its support structure, only one OTP point distributed on the locator is taken as the actual measurement datum. And the position of the above OTP point represents the spatial position of the entire end effector within the locating unit.

③ For the motion process of each axis, the feed direction is parallel to the coordinate axis, and the coordinate axes in the tooling coordinate system would not rotate relative to each other during the locating process.

④ For the locating process of flexible tooling, it is not affected by the practical assembly conditions/environment. It is assumed that the external disturbances, such as time-varying state parameters, have no influence on the positioning accuracy of tooling at this pre-compensation stage. Because of the existence of the above factors, the error transfer exhibits a non-liner relationship, and the accurate mathematical modeling for the changes of positioning accuracy is always difficult to accomplish.

⑤ In each motion direction, the installation error of the motion module is a variable value within the movement stroke, such as in the ball screw system and the guiding drive system. However, it is considered to be a fixed value for the convenience of subsequent modeling.

In summary, based on the actual error status and assembly deformation, the above assumptions would simplify the error transfer modeling process and enhance its modeling feasibility. The detailed modeling process is taken as the geometric error propagation along three motion axis, and the deformation error accumulation at the end locator, as shown below.

4.4.2　Positioning Error Modeling for One Single Motion Axis

For the positioning error modeling on a single motion axis, the error in the x direction is taken as a modeling example.

4.4.2.1　Establishment of Local Coordinate System

For the reference point P that located on the end locating effector, in order to model the positioning error variation ΔP, three local tooling coordinate systems are established firstly, as shown in Figure 4-7.

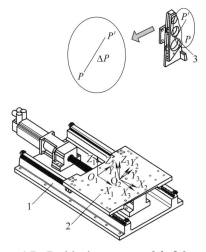

Figure 4-7　Positioning error model of the x axis

1—Pedestal of the x direction; 2—Connecting plate; 3—End locating effector

① A fixed rectangular coordinate system, denoted as D_1 (O_1, X_1, Y_1, Z_1), is set up at the symmetrical center point of the x-direction pedestal. The direction of its each coordinate axis coincides with the motion direction of the locating unit. ② Considering the influence of the motion error and installation error of the comprised removable modules according to ΔP, the coordinate system D_2 (O_2, X_2, Y_2, Z_2) is established at the center of the connection plate. ③ Another coordinate system, D_3 (O_3, X_3, Y_3, Z_3), is also established at the center of the connection plate. D_3 is an auxiliary tooling coordinate system used for error modeling. The original points of D_2 and D_3 overlap with each other. However, the axis direction of D_3 is parallel to the fixed axis direction of D_1 coordinate system, and the direction would not rotate in the entire practical motion process. Only if there is no installation orientation error in the guiding rail system, D_2 and D_3 coincide with each other. And this is the difference between D_2 and D_3. In the actual positioning process, due to the geometric errors of the motion modules, there is a certain angle between the two coordinate systems.

4.4.2.2 Installation Orientation Error Modeling for Guiding Drive System

In order to understand the influence of installation orientation error in the guiding drive system on positioning reference point P, the error transfer relationship should be analyzed firstly under the assumption that there is no installation error, i.e., the ideal situation. Suppose the coordinates of position vector OP for reference point P are (x, y, z) in D_2 and D_3 coordinate systems. Since the coordinate axis of X_2 is parallel to X_1, considering the positional relationship between D_1 and D_2, the position of P in D_1 is $(x + x_0, y, z)$. In practical locating process, considering the existence of straightness error items in horizontal/vertical planes for one guiding rail, or straightness error items of the guiding rails relative to the reference axis, and parallelism error items between different guiding rails, the vector OP would have a translation of Δy and Δz along the axis directions of Y_1 and Z_1. In addition, there also exists a rotation angel of α, β, γ around X_1, Y_1, Z_1 axes. As a result, the actual position of point P will deviate from its theory position, moving to a new position, i.e., (x', y', z'). And the corresponding position vector OP' could be expressed in Equation (4-1). Where $R(i, j)$ is the rotation matrix, which standing for the angel j around the i axes, and $i= X_1, Y_1, Z_1$, $j=\alpha, \beta, \gamma$.

$$OP' = R(X_3,\alpha)R(Y_3,\beta)R(Z_3,\gamma)OP = E_1 OP \tag{4-1}$$

Since all of the angle displacement errors are small quantities, according to the differential rotation transformation theory, the rotation transformation matrix E_1 can be expressed with Equation (4-2). Where the positive and negative signs of α, β, γ are determined by the right hand rule. It can be proved that the calculated results have no relationship with the rotation order around the three motion axes, i.e. X_3, Y_3, and Z_3.

$$E_1 = \begin{bmatrix} 1 & -\gamma & \beta \\ \gamma & 1 & -\alpha \\ -\beta & \alpha & 1 \end{bmatrix} \tag{4-2}$$

Given the translation error of coordinate system D_2 is not taken into consideration, when the connecting plate moves to the point x_0 from its original position, the actual coordinates of point P in the D_l coordinate system would be changed to $(x'+x_0, y', z')$. Then, the position error of point P could be calculated based on the installation orientation error of the guiding drive system, as shown in Equation (4-3).

$$\begin{bmatrix} x'+x_0 \\ y' \\ z' \end{bmatrix} - \begin{bmatrix} x+x_0 \\ y \\ z \end{bmatrix} = \begin{bmatrix} x' \\ y' \\ z' \end{bmatrix} - \begin{bmatrix} x \\ y \\ z \end{bmatrix} = OP' - OP = (E_1 - I)OP = \begin{bmatrix} \Delta x_r \\ \Delta y_r \\ \Delta z_r \end{bmatrix} = \begin{bmatrix} z\beta - y\gamma \\ x\gamma - z\alpha \\ y\alpha - x\beta \end{bmatrix}$$

$$\tag{4-3}$$

4.4.2.3　Installation Posture Error Modeling for Ball Screw System

The leading precision of ball screw system, is defined as the positioning error of the center point for the ball screw axis. Its calculation usually needs the measurement of the distance from the original point, such as the center point of the screw in length, to the required motion position. Given the feedback measurement error of grating ruler is not considered for each motion axis, the points that distributing on the entire ball screw axis have different position deviations along the vertical direction and horizontal directions, within the whole range of the motion stroke. And this posture change could explain why the installation posture error occurs. Then the positioning error of locating unit would be affected by the posture variation in motion process. To be more specific, it mainly includes the angle error in the left and right directions, i.e., the horizontal posture error, and the error in the vertical direction, as shown in Figure 4-8.

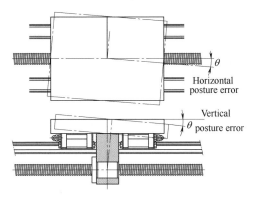

Figure 4-8　Operation posture variation for ball screw drive system

In the D_3 coordinate system, assuming that the vertical (or horizontal) distance between the positioning reference point P and the center of the ball screw is L, the angle between the position vector O_2P and the coordinate axes of the D_3 coordinate system is (α', β', γ'), with the posture error in vertical/horizontal direction is denoted as θ. Then the positioning error of the point P caused by the installation posture error can be expressed as:

$$A = L\sin\theta \tag{4-4}$$

With the decomposition of vector A along the direction of each coordinate axes, then for the error point P, caused by the vertical/horizontal posture error, can be expressed as:

$$A_V = L\sin\theta \begin{bmatrix} \sin\beta' \\ \sin\alpha' \\ 0 \end{bmatrix} \tag{4-5}$$

$$A_H = L\sin\theta \begin{bmatrix} \sin\gamma' \\ 0 \\ \sin\alpha' \end{bmatrix} \tag{4-6}$$

Combining the above two equations, the final positioning error of point P that caused by operational posture variation, due to the lead precision of ball screw system, can be obtained, shown as:

$$A_\Sigma = A_V + A_H = L\sin\theta \begin{bmatrix} \sin\gamma' + \sin\beta' \\ \sin\alpha' \\ \sin\alpha' \end{bmatrix} \tag{4-7}$$

4.4.2.4　Axial Feed Error Modeling for Ball Screw System

Assuming there is a motion input value x_0 along the x direction in the control system of the locating unit, with a comprehensive consideration of axial feed error Δx_t and the motion posture variation A_Σ due to the leading accuracy of ball screw system, the specific comprised components for the positioning error of point $P(x,y,z)$ can be expressed as:

$$\Delta X_P = \begin{bmatrix} \Delta x \\ \Delta y \\ \Delta z \end{bmatrix} = \begin{bmatrix} \Delta x_r \\ \Delta y_r \\ \Delta z_r \end{bmatrix} + L\sin\theta \begin{bmatrix} \sin\gamma' + \sin\beta' \\ \sin\alpha' \\ \sin\alpha' \end{bmatrix} + \begin{bmatrix} \Delta x_t \\ 0 \\ 0 \end{bmatrix} \tag{4-8}$$

According to Equation (4-1)~(4-8), assuming there are other input feed values y_0 and z_0 in the control system of the locating unit along the y and z directions, then the positioning error items of point $P(x,y,z)$ can be expressed as:

$$\Delta Y_P = \begin{bmatrix} \Delta x \\ \Delta y \\ \Delta z \end{bmatrix} = \begin{bmatrix} \Delta x_r \\ \Delta y_r \\ \Delta z_r \end{bmatrix} + L\sin\theta \begin{bmatrix} \sin\beta' \\ \sin\alpha' + \sin\gamma' \\ \sin\beta' \end{bmatrix} + \begin{bmatrix} 0 \\ \Delta y_t \\ 0 \end{bmatrix} \tag{4-9}$$

$$\Delta Z_P = \begin{bmatrix} \Delta x \\ \Delta y \\ \Delta z \end{bmatrix} = \begin{bmatrix} \Delta x_r \\ \Delta y_r \\ \Delta z_r \end{bmatrix} + L\sin\theta \begin{bmatrix} \sin\gamma' \\ \sin\gamma' \\ \sin\alpha' + \sin\beta' \end{bmatrix} + \begin{bmatrix} 0 \\ 0 \\ \Delta z_t \end{bmatrix} \tag{4-10}$$

4.4.3　Positioning Error Modeling for Spatial Motion Axes

According to the first assumption in Section 4.1.1, it's known that the total spatial positioning error of the reference point P distributed on the end locators, can be taken as the sum of the linear displacement errors generated in each motion direction. For the calculation of integrated accumulation error, according to Equation (4-1)~(4-10), their

Digital Assembly Coordination and Quality Controlling Technology
for Aeronautical Thin-walled Structures

extreme values is considered, and the absolute values of each motion direction are summed together to obtain the final positioning error, as shown in Equation (4-11).

$$[\Delta X, \Delta Y, \Delta Z]^{\mathrm{T}} = \Delta \boldsymbol{X}_P + \Delta \boldsymbol{Y}_P + \Delta \boldsymbol{Z}_P \tag{4-11}$$

Then, the deviation between the actual spatial position and the theoretical position of the positioning reference feature P is:

$$\Delta P_G = \sqrt{\Delta X^2 + \Delta Y^2 + \Delta Z^2} \tag{4-12}$$

4.4.4 Deformation Error for the End Locating Effectors

In the process of flexible assembly, the tooling will be deformed under the action of assembly force and other extra loads, as shown in Figure 4-5. It's mentioned that, for the actual spatial position of the end locaters, it can be measured with the assistance of the laser tracker. According to the second assumption in Section 4.1.1, the measured results have a function of verifying the correctness of the calculation results. And the assembly force and strain can be sensed in real time with the three-component force sensor and strain gauge. As a result, the measured deformed position ΔP_D of the end locating effectors can be recorded and then compensated to the calculated positioning error ΔP_G with geometric error transfer analysis. As modeling the deformation error, by constructing the stiffness matrix of the given end locating effectors, the relationship between assembly force and assembly deformation can be gained according to the finite element simulation method. With another perspective, the locating unit can be taken as a freely supported beam. This error could also be obtained with the help of deformation calculation based on the theory of mechanics of materials.

$$\Delta P = \Delta P_G + \Delta P_D \tag{4-13}$$

4.4.5 Experiment Verification

At practical assembly site, firstly, with the help of laser tracker (API track III), an assembly measuring field for the flexible tooling system is built. The measuring datum consists of ① four TB (tooling ball) points that distributed at the corner of the adjustable support pedestal, and ② sixteen ERS (enhanced reference system) points, as shown in Figure 4-9. The first three TB points are taken as the basis for establishing the TCS (tooling coordinate system), and the fourth TB point has a function of checking and adjusting the built reference system. ERS points have a function of establishing the local TCS. Where, $OXYZ$ stands for the theoretical TCS, and $O'X'Y'Z'$ represents the practical TCS. In the coordinate system, the actual spatial position (x, y, z) of the end locaters can be obtained.

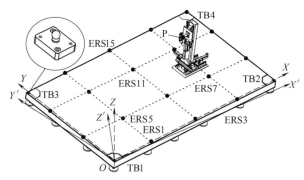

Figure 4-9   Coordinate system of the flexible assembly tooling

Due to the flexible tooling system has a compact structure, the locating unit needs to be installed layer by layer. In order to ensure the spatial accuracy of the motion modular on each layer, higher requirements are put forward for the position accuracy of their comprised parts, such as ball screw drive system and guiding drive system. With the premise that all the parts and linear motion systems of the locating unit have been checked and passed the quality inspection, combing with the practical manufacturing and installation process, the adjustment and installation works for each motion direction could be carried out. Then the actual geometric error status can be recorded. Considering the actual operation ability, it is found that the installation orientation error of guiding drive system for each motion axis can be taken as the same value, of 0.05mm/1000mm, and the installation posture error of ball screw system is considered to be 10 seconds, with the help of measurement equipment that listed in Figure 4-3. According to the precision characteristics and error distribution of guiding drive system, and the installation instructions described in the technical manual book provided by Bosch Company (Germany), the allowable value of height deviation and parallel deviation for different nominal models are determined, as shown in Figure 4-10. Then, with the detailed calculation method shown in Table 4-2, the installation orientation angle can be obtained for each motion direction.

Table 4-2   Description on the installation orientation error for guiding drive system

Unit: (°)

Error item	Symbol	Causation	Function size	Error value
Inclination angle	α	Transverse height deviation between rails, S_1	Distance between rails, a	S_1/a
Elevation angle	β	Longitudinal height deviation between rails, S_2	Distance between slides, b	S_2/b
Deflection angle	γ	Parallelism deviation between rails, P_1	Distance between locating edges, c	P_1/c

Digital Assembly Coordination and Quality Controlling Technology
for Aeronautical Thin-walled Structures

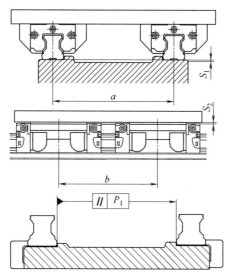

Figure 4-10   Installation orientation error of the guiding drive system

Within the whole motion stroke, the axial feed error report of ball screw drive system is provided by Bosch Company. The motion direction of x-axis is taken as an example, as shown in Table 4-3. According to the structural characteristics of ball screw drive system and the error values for the four given error items, the motion error of other desired position within the motion stroke range, can be gained with the method of linear interpolation.

Table 4-3   Error report of ball screw drive system

Motion stroke/mm	Stroke error/μm
0.00	0.0
50.00	−5.1
100.00	−9.6
150.00	−13.2
200.00	−16.3
250.00	−20.7
300.00	−25.9

As installing and debugging of the flexible assembly system, for the positions of the end effectors, it would be checked and adjusted to fit the precision requirements comparing with practical OTP measurement results[37]. With the real-time assistance of laser tracker, five main steps are designed in the positioning accuracy guarantee scheme, as shown in Figure 4-11.

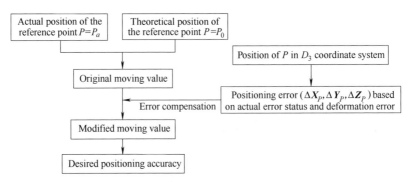

**Figure 4-11   Positioning accuracy guarantee scheme for the
locating unit with laser tracker**

Step 1   Measure the theoretical position of reference point P in tooling coordinate system in CATIA software environment, and then a coordinate value is recorded as $P_0 = (x_0, y_0, z_0)$.

Step 2   In the actual assembly environment, measure the position of point P ten times with a laser tracker at the original point along the $X/Y/Z$ axis in the TCS. And then the average coordinate value is denoted as $P_a = (x_a, y_a, z_a)$.

Step 3   Calculate the difference between P_0 and P_a. Their deviation result $(\Delta X_0, \Delta Y_0, \Delta Z_0)$ can be taken as the original moving value to be input in the control system.

Step 4   According to the mathematic model of positioning error ΔP_G and deformation position ΔP_D for the locating unit, calculate the comprehensive positioning error ΔP between theoretical and actual positions at reference point P on the end locator, where $\Delta P = (\Delta X_P, \Delta Y_P, \Delta Z_P)$. By adding $(\Delta X_0, \Delta Y_0, \Delta Z_0)$ and ΔP together, and then a modified moving value can be obtained. It is mentioned that the above mathematic model is constructed based on the actual error status of ball screw drive system, guiding drive system, and other motion parts, with error transfer mechanism analysis.

Step 5   On the basis of the modified motion value, move the locating unit from original point of the $X/Y/Z$ axis to the desired locating position. The measurement operations for the position of reference point P could also be taken with laser tracker. Then the positioning accuracy can be checked whether it can meet the requirement of 0.05mm.

For the given assembly work of wing flap components at four different assembly stations, the actual positioning precision of one locating unit is analyzed and recorded following the previous 5 steps, as shown in Table 4-4.

Table 4-4　Positioning error pre-compensation considering actual
error status and deformation error

Assembly station	Theoretical coordinates of reference point P/mm	Measured coordinates of reference point P/mm	Coupling error considering actual geometric status and deformation error/mm	Modified feed value/mm	Measured positioning error/mm
1#	3842.414,1721.038, 1082.672	3923.447,1814.802, 1070.619	0.154,0.354,0.228	81.187,94.118, 11.825	0.023,−0.039,−0.021 ΔP=0.049
2#	4762.844,1836.937, 1143.876	4849.859,1885.471, 1159.909	0.187,−0.294,0.213	87.202,48.240, 16.246	−0.021,0.038,−0.024 ΔP=0.050
3#	3794.205,1654.821, 1209.737	3919.180,1683.169, 1265.683	−0.246,0.303,0.182	124.729,28.651, 56.128	−0.016,−0.041,−0.013 ΔP=0.050
4#	3054.231,1804.324, 1194.378	3179.038,1882.293, 1251.003	−0.213,0.252,−0.189	124.594,78.221, 56.436	0.024,−0.037,−0.008 ΔP=0.045

4.4.6　Discussion

With the analysis of the above four groups of experiment data, considering actual geometric error status and deformation errors, after the pre-compensation with the error coupling operation, the positioning error of the locating unit is limited to within 0.05mm. The compensated error can basically meet the assembly accuracy requirements for aircraft parts in practical engineering processes. This indicates the effectiveness of pre-compensation strategy, with the above pre-compensation stage also being taking as the first step for precise modeling of positioning accuracy.

However, the above compensation strategy is more suitable for the installation stage of each locating unit. It is mentioned that, for the practical assembly process, there are many time-varying parameters influencing the positioning accuracy of the tooling. The practical assembly conditions and the non-linear coupling relationship cannot be reflected in the above error transfer mechanism. As a result, only the five assumptions and the linear accumulation relationships are analyzed before. Considering there are enhancement accuracy requirements for flexible assembly tooling in the new generation aircrafts, the influence of the time-varying parameters that mentioned in the Introduction Section on the actual positioning error, cannot be ignored any more. The dynamic compensation based on real-time measurement for the entire tooling system should be further researched.

4.5　Parameter Identification for Tooling and Locating Motion Model with Measured Sample Data

The structure of the locating unit mentioned in the Introduction Section can be taken as a serial locating motion mechanism[38]. Intuitively, its final positioning accuracy is determined by each comprised liner motion system along three axes. However, considering the actual assembly conditions, such as the actual status of tooling/product/ environment, the assembly deformation, and the changing loads on end locators, the motion model of the locating unit is not constant. The influence principle of time-varying state parameters on the positioning accuracy of the tooling should be further studied, with the purpose of obtaining the precise input and output relationships and building an accurate locating motion model. However, there is a significant phenomenon indicates that their precision relationships are nonlinear and strong coupling. Due to the practical assembly conditions, there is a cassette between the motion input and the motion output, and the relationships are not always explicit. For example, the accurate error transfer principle cannot be gained under the above situations. To find the correlation relationships, and to identify the motion control parameters of the actual motion model for the locating unit, parameter identification based on measurement data for locating motion model is an efficient method. Then the transfer process of the inputs and the compensation of motion parameters would be understandable.

4.5.1 Parameter Identification Process for Locating Motion Model

Due to the characteristic of multiple-input and multiple-output(MIMO) system, for the positioning accuracy of the locating unit, it is affected by the practical time varying constraint conditions. It is known that the accurate theoretical motion model cannot be established easily. This requires the identification and modeling of ① motion models with measuring devices, and ② methods of function approximation, such as multi-layer neural network, fuzzy neural network, and other intelligent methods[39]. For the traditional learning algorithms, such as BP (back propagation) algorithm, the gradient descent method is mostly adopted. The network structure needs to be tested according to manual experience, because ① many parameters to be found and optimized, ② a slow convergence speed, and ③ a long training process. In addition, the above algorithms are easy to fall into local minimum point and sensitive to the learning rate setting. However, the SLFN (single-hidden layer feedforward neural-network) method is widely used in parameter identification, due to the good learning ability. And there is an ELM (extreme learning machine) learning algorithm for SLFN. ELM can generate the random connection weights between the neurons in input layer and hidden layer, and the threshold for the hidden layer neurons. For the weight numerical values, adjustments are not required during training process. Only the number of hidden layer neurons needs to be set, and then the unique optimal solution can be obtained. In summary, the above solution has the advantages of fast learning and good generalization. For the locating unit showed in Figure 4-1, a solution of model identification based on assembly statistical data is suitable for modeling the input and output coupling relationships[40,41]. Based on the practical measurement data, three main steps are contained: parameter identification for motion model of the locating unit, SLFN modeling, and parameter learning algorithm based on ELM, as shown in Figure 4-12.

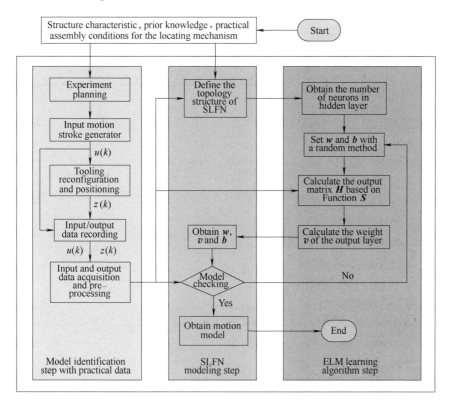

Figure 4-12   Parameter identification process for positioning motion model

The detailed explanations of the above three steps can be shown as follows.

Step 1   According to the structure characteristic, prior knowledge of the locating/ assembly process, and practical assembly conditions for the locating mechanism, design experimental scheme firstly. With the input motion stroke generator, the motion data is fed into the drive system in each direction. By measuring the output value, i.e., the positioning error of the locating unit, these input/output data is pre-processed and analyzed with the purpose of determining the topology structure of the SLFN model. It is mentioned that the measurement data is divided into two portions: the training samples (for identifying the parameters of SLFN prediction model) and the testing samples (for verifying the effectiveness of the model).

Step 2   Determine the detailed structure of SLFN prediction model. With the solutions of ① determining the number of neurons in hidden layer, ② setting an initial weight w that connecting the input layer and hidden layer, ③ setting an initial b which connects the hidden layer and output layer, and ④ selecting an infinitly differentiable function S, then the output matrix H of the hidden layer and the weight v of the output layer, can be calculated based on the training data and the ELM method. It is mentioned that the function S is considered as an activation function of hidden layer neurons.

Step 3 Model validation with the obtained model structure parameters of w, v, and b, and the experiment data, i.e., testing samples. If the built model is checked to be qualified, the parameter identification work for positioning motion model can be finished. Otherwise, the parameters w and b should be re-generated with different values.

4.5.2　Topology Structure of SLFN Prediction Model

For the SLFN prediction model, three partitions are mainly contained: the input layer, the hidden layer, and the output layer. The neuron nodes between each layer are fully connected, as shown in Figure 4-13.

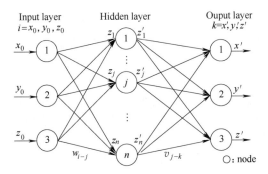

Figure 4-13　Topology structure of the SLFN error prediction model

To be more specific, for the SLFN model that used for predicting the output positioning error along three motion axes, there are three nodes in the input layer, denoted as x_0, y_0, and z_0. They represent the given motion values of each axis, i.e., x, y, and z, respectively. Correspondingly, three neuron nodes are also contained in the output layer, i.e. x', y', and z', representing the final positioning errors for the locating unit. For the hidden layer, n neuron nodes are contained, which stand for the jth input z_j and the jth output z_l'. For the link weights between different neuron nodes of different layers, w_{i-j} represents the weight between the input layer and the hidden layer, v_{j-k} represents the weight between the hidden layer and the output layer, and b_j represents the threshold of the jth neuron for the hidden layer, where, $j=1, 2, \cdots, n$.

For the topology structure of the SLFN model, the jth input z_j can be taken as the weighted sum of all the input nodes, as follows:

$$z_j = \sum_{i=1}^{3} w_{i-j} x_i \tag{4-14}$$

The jth output z_j' can be gained with a function S based on z_j, as follows:

$$z_j' = S(z_j + b_j) = 1/[1 + e^{-(z_j + b_j)}] \tag{4-15}$$

Following the same approach, for the output layer, the calculation of the kth output variable x_k', can be expressed as:

$$x_k' = \sum_{j=1}^{3} v_{j-k} z_j' \tag{4-16}$$

Based on the above three equations, the prediction model for positioning error based on SLFN can be obtained, where only the input variable x_i, the output variable x_k', and the constant variable w, v, and b are contained in the model.

$$x' = v^{\mathrm{T}} z' = \begin{bmatrix} \sum_{j=1}^{3} v_{j-1} / [1 + e^{-\left(\sum_{i=1}^{3} w_{i-j} x_i + b_j\right)}] \\[2mm] \sum_{j=1}^{3} v_{j-2} / [1 + e^{-\left(\sum_{i=1}^{3} w_{i-j} x_i + b_j\right)}] \\[2mm] \sum_{j=1}^{3} v_{j-3} / [1 + e^{-\left(\sum_{i=1}^{3} w_{i-j} x_i + b_j\right)}] \end{bmatrix}_{3 \times 1} \tag{4-17}$$

4.5.3 Parameter Identification on SLFN Locating Motion Model Based on ELM

Under the instruction of the **Step 2** in Section 4.5.1, for the locating unit, given there is an input motion matrix containing P training samples, denoted as $X \in R^{3 \times P}$, then the pth training data sample can be expressed as:

$$X_p = [x_{0p} \quad y_{0p} \quad z_{0p}]_{1 \times 3}^{\mathrm{T}} \tag{4-18}$$

And for the corresponding output matrix, representing the output positioning error, the pth training data sample can be expressed as:

$$Y_p = [x_p' \quad y_p' \quad z_p']_{1 \times 3}^{\mathrm{T}} \tag{4-19}$$

Where $Y_p \in R^{3 \times P}$.

Supposing there is an output result from SLFN prediction model, it can be expressed as:

$$T = [t_{x-p} \quad t_{y-p} \quad t_{z-p}]^{\mathrm{T}}_{1\times3} \qquad (4\text{-}20)$$

With Equation (4-17)~(4-20), the predicted positioning error can be gained, and there is:

$$t_p = \begin{bmatrix} t_{x-p} \\ t_{y-p} \\ t_{z-p} \end{bmatrix}_{3\times1} = T^{\mathrm{T}} = \begin{bmatrix} \sum\limits_{j=1}^{3} v_{j-1} f(w_j X_p + b_j) \\ \sum\limits_{j=1}^{3} v_{j-2} f(w_j X_p + b_j) \\ \sum\limits_{j=1}^{3} v_{j-3} f(w_j X_p + b_j) \end{bmatrix}_{3\times1} = Hv \qquad (4\text{-}21)$$

In the topology structure of the SLFN prediction model, H stands for the output matrix of the hidden layer. Under the guidance of ELM parameter learning algorithm, and a random parameter value of w and b, it is known that if the number of training samples is equal to the number of neuron nodes in hidden layer, then the output results of SLFN prediction model can match the training samples, with zero tolerance. However, the neurons number is usually less than P. In this situation, because the function S, as mentioned in Equation (4-15), is infinite differentiable, then not all of the parameters in SLFN prediction model should be adjusted. The parameter of w and b can be set randomly before the data training process, and they would remain constant in the training procedure. As a result, the weight v_{j-k} between the hidden layer and the output layer can be gained, by means of solving the least square solution of the previous equation set, as:

$$\min_{v} \| Hv - T^{\mathrm{T}} \| \qquad (4\text{-}22)$$

The optimization solution of the weight is:

$$\hat{v} = H^{+} T^{\mathrm{T}} \qquad (4\text{-}23)$$

Where H^{+} represents the generalized Moore-Penrose inverse matrix of H.

4.5.4 Verification and Analysis of Tooling Locating Motion Model

For the positioning error of the locating unit, under the instruction of **Step 3** in Section 4.5.1, the parameter identification work for tooling SLFN error prediction model is verified. In the assembly measurement setup field, the grating ruler (Renishaw) is used to record the feed position data of each motion axis, while the laser tracker (API III) is used to record the actual spatial position of the end locating effectors that mounted on the tooling. It is mentioned that the reference coordinate system is shown in Figure 4-9. All the data samples can be acquired in practical assembly site, where a lot of time-varying parameters are contained. The detailed experiment scheme for verifying the SLFN model

Digital Assembly Coordination and Quality Controlling Technology
for Aeronautical Thin-walled Structures

can be described as follows. Firstly, within the range of locating motion stroke for each motion axis, i.e. 300mm, a group of 150 data samples for the motion feed data along each axis and the corresponding spatial position is recorded, totaling 450 data samples are contained for the entire locating unit along three directions. The data acquisition interval is equal to 2mm. These data are also taken as the training samples of SLFN prediction model. Secondly, as the three motion axes of the locating unit are driven at the same time in practical assembly site, a group of 30 data samples that are evenly distributed within three motion stroke is also recorded. Different from the previous situation, the data collection interval is 10mm. And this kind of data is taken as the testing samples for the verification of the SLFN prediction model.

For the individual x, y, and z motion axis, the measured training data set that standing for the positioning error of the end locators is shown in Table 4-5~4-7, respectively.

Table 4-5　The measured training data set for x motion axis

No.	x input /mm	x output /mm	y output /mm	z output /mm	No.	x input /mm	x output /mm	y output /mm	z output /mm
1#	2	2.003	0.001	0.005	76#	152	152.071	0.005	−0.021
2#	4	4.006	−0.003	−0.003	77#	154	154.072	0.003	−0.017
3#	6	6.004	−0.001	−0.001	78#	156	156.069	0.012	−0.020
4#	8	8.004	0.003	0.002	...	......	......	......	......
5#	10	10.010	−0.005	0.001	136#	272	272.124	0.020	−0.028
...	......	......	......	......	137#	274	274.125	0.018	−0.023
73#	146	146.063	0.014	−0.007	138#	276	276.130	0.015	−0.028
74#	148	148.073	0.004	−0.010	139#	278	278.130	0.017	−0.025
75#	150	150.072	0.004	−0.004	...	......	......	......	......

Table 4-6　The measured training data set for y motion axis

No.	y input /mm	x output /mm	y output /mm	z output /mm	No.	y input /mm	x output /mm	y output /mm	z output /mm
1#	2	0.007	1.990	−0.011	76#	152	0.010	152.040	−0.023
2#	4	0.012	3.980	−0.017	77#	154	0.009	154.046	−0.031
3#	6	0.014	5.970	0.009	78#	156	0.008	156.045	−0.026
4#	8	0.012	7.982	−0.080	...	......	......	......	......
5#	10	0.007	9.992	−0.009	136#	272	0.022	272.087	−0.039
...	......	......	......	......	137#	274	0.025	274.085	−0.036
73#	146	0.014	146.032	−0.025	138#	276	0.021	276.090	−0.036
74#	148	0.008	148.042	−0.032	139#	278	0.025	278.085	−0.040
75#	150	0.017	150.031	−0.029	...	......	......	......	......

Table 4-7   The measured training data set for z motion axis

No.	z input /mm	x output /mm	y output /mm	z output /mm	No.	z input /mm	x output /mm	y output /mm	z output /mm
1#	2	−0.006	0.007	1.948	76#	152	0.037	0.0104	151.967
2#	4	−0.006	0.008	3.948	77#	154	0.038	0.0098	153.965
3#	6	−0.005	0.010	5.947	78#	156	0.038	0.011	155.963
4#	8	−0.007	0.009	7.950	...	......	......	......	......
5#	10	−0.007	0.009	9.950	136#	272	0.060	0.0148	271.960
...	......	......	......	......	137#	274	0.059	0.0165	273.964
73#	146	0.035	0.011	145.966	138#	276	0.059	0.0191	275.964
74#	148	0.036	0.011	147.963	139#	278	0.059	0.0195	277.965
75#	150	0.038	0.010	149.965	...	......	......	......	......

When the three motion axis are driven at the same time, the measured testing data set for the positioning error of the end locators is shown in Table 4-8.

Table 4-8   The measured testing data set for three motion axis

No.	Three input /mm	x output /mm	y output /mm	z output /mm	No.	Three input /mm	x output /mm	y output /mm	z output /mm
1#	10/10/10	10.007	10.011	9.931	16#	160/160/160	160.117	160.082	159.944
2#	20/20/20	20.018	20.007	19.934	17#	170/170/170	170.125	170.085	169.943
3#	30/30/30	30.026	30.013	29.937	18#	180/180/180	180.131	180.090	179.945
4#	40/40/40	40.028	40.026	39.938	...	......	......	......	......
5#	50/50/50	50.036	50.029	49.939	24#	240/240/240	240.184	240.123	239.942
...	......	......	......	......	25#	250/250/250	250.196	250.127	249.946
13#	130/130/130	130.088	130.073	129.949	26#	260/260/260	260.200	260.141	259.942
14#	140/140/140	140.099	140.074	139.949	27#	270/270/270	270.214	270.144	269.941
15#	150/150/150	150.104	150.083	149.951	...	......	......	......	......

Due to the number of training samples for all three motion axes is 450, i.e., $P = 450$. Regarding to the number of neuron nodes in hidden layer, its value can also be set to 450 for the normalization processing reason, i.e., $n = 450$. Given an initial setting of the parameters for w and b, the definition of function S, and the calculation of the output matrix H for the hidden layer based on the training/testing data samples, the parameter identification results based on ELM learning algorithm for tooling locating motion model is shown as follows.

$$b = [0.452 \quad 0.984 \quad \cdots \quad 0.329]^{\mathrm{T}}_{1 \times 450} \tag{4-24}$$

$$w = \begin{bmatrix} 0.674 & 0.988 & -0.645 \\ 0.346 & -0.239 & 0.283 \\ \vdots & \vdots & \vdots \\ 0.450 & 0.582 & 0.845 \end{bmatrix}^{\mathrm{T}}_{450 \times 3} \tag{4-25}$$

$$v = \begin{bmatrix} 34854.433 & 78043.898 & 63381.622 \\ 17927.825 & -46243.840 & 45325.457 \\ \vdots & \vdots & \vdots \\ -9344.741 & 58652.084 & 42529.505 \end{bmatrix}^{\mathrm{T}}_{450 \times 3}$$ (4-26)

With the obtained parameters, w, v, and b, and the 450 training data samples, the effectiveness of SLFN prediction model can be validated. With the de-normalization operations of the output results, the prediction values and the corresponding errors for the training sample sets are analyzed. The mean square deviation of the fitting ability for the SLFN model is 0.005mm, with the consideration of training data samples. Then, for the 30 testing samples, their prediction error along each motion direction is analyzed, as shown in Figure 4-14. It is also found that the mean square deviation of the prediction

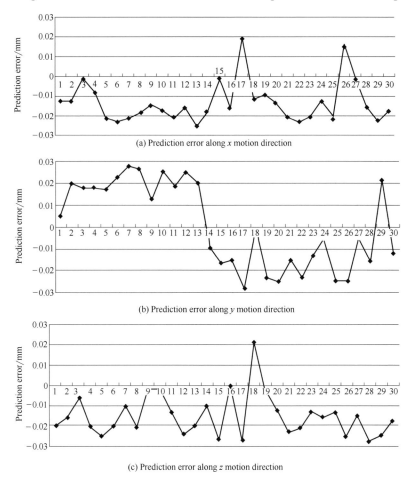

(a) Prediction error along x motion direction

(b) Prediction error along y motion direction

(c) Prediction error along z motion direction

Figure 4-14　Prediction error for the testing samples along each motion direction

ability for the SLFN model is 0.0029mm, the maximum mean deviation is 0.0227mm at the 22nd testing sample.

With the above modeling for parameter identification and analysis of the measured data in practical assembly site, a conclusion can be obtained: the calculation/prediction results of the SLFN prediction model show high consistency with the measured positioning errors in the practical assembly process. With the experimental design and development of the positioning motion model, the methodology's effectiveness has been verified. The model has a higher prediction ability, leading to a good foundation for the accurate positioning error guarantee of the locating unit.

4.6 Accurate Compensation Stage Combining Mechanism Analysis and Measurement Data

In this chapter, the relationship between the mechanism model mentioned in Section 4.4 and the SLFN prediction data model in Section 4.5 is analyzed. A new positioning error guarantee model is proposed, with the purpose of realizing the accurate compensation for the final positioning error by combining the advantages of mechanism model and data model.

4.6.1 Logical Relationship Between Two Compensation Stages for Positioning Error Guarantee

For the mechanism model of positioning error at the pre-compensation stage, considering the locating unit's comprised geometric error items and assembly deformation conditions, the error transfer and coupling principle are analyzed. However, due to the complex variables in practical assembly site, five assumptions are stated to decrease the modeling difficulty. This simplified solution would lead to an inaccurate explanation of the correlation between the input and output results, because ① the disturbances of practical time-varying parameters, and ② the assumption of error accumulation linear relationship. Then, the five compensation steps mentioned in positioning accuracy guarantee scheme for the locating unit are carried out, as shown in Figure 4-11, it is found that there is a notable difference between the measured positions and the theoretical coordinates, indicating the practical situation cannot be reflected. With the pre-compensation strategy based on actual error status and deformation analysis, the average positioning error for the four assembly stations is 0.0485mm. However, the acquired error locates at the edge of required tolerance range. Because the measurement error of the laser tracker is not taken into account, the compensated results have a great deviation exceeding risk.

For the data model of the locating unit, i.e., the SLFN prediction model, according to the uniform sampling of the input motion values and the measured positioning errors, it is found the model has an excellent prediction ability, with a clear non-linear coupling relationship. With this data model, not only the relevant correlation factors with the positioning accuracy can be expressed, but also the unknown parts of the model. Where, the prior assembly knowledge is not required. And the key parameters of the model, such as w, v, and b, are obtained to understand the truth of positioning error guarantee with data model. Although the prediction results is in accordance with the practical measurement results, there are several disadvantages. For example, there is a poor interpretability for data models, causing the detailed error transfer principle, and the causal relationship between input motion driven values and output positioning errors cannot be gained. Secondly, the prediction accuracy is easily affected by the quantity and quality of sample data, making the stability of prediction results cannot be guaranteed efficiently. Furthermore, the data model needs to be updated and maintained regularly, with the purpose of adapting to new assembly scenes and enhance the application scope of the model.

In order to solve the above problems for the two models, and make the models more convenient for practical assembly site, a comprehensive combination of mechanism model and SLFN prediction model is proposed, i.e., semi-mechanism model. With the accurate compensation operation, the output results of the new model is taken as the final positioning errors. The specific ideas are shown in Figure 4-15.

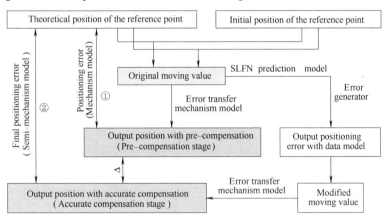

Figure 4-15　Comprehensive combination of mechanism model and data model
(See the color illustration)

There are two closed cycles for the accurate positioning error guarantee process. The first one describes the error compensation method based on error transfer mechanism model, as detailed description in Figure 4-11. Then, the accurate compensation stage with mechanism analysis and measurement data, can be shown by the second cycle, i.e., the semi-mechanism modeling. With the difference value between the theoretical and initial position of the reference point P, an original moving value is input into the control system. Then based on this new SLFN prediction model in this section, prediction results of the corresponding positioning error is gained with the help of the training and testing data samples. This prediction process plays the role of error generator. After the above error items in each motion direction are compensated to the original moving value, the modified motion value is obtained and input to the error transfer mechanism model. For the locating unit, comparing the output position with the above accurate compensation and the theoretical position, the methodology's effectiveness can be gained.

4.6.2 Error Generator based on Error Measurement Data

For the same locating unit, the prediction work of positioning error is done, according to the modeling and analysis steps that already mentioned in Figure 4-12 in the last setion. Through the operation of practical error data collection and processing, an error generator is built. Following the same method, for the SLFN prediction model, there are three neuron nodes each in the input layer and output layer. Within the range of motion stroke for each axis and the simultaneous motion of three axes, a total of 168 training data samples for the entire locating unit and 20 testing data samples for the simultaneous motion along three motion directions are collected. Considering the actual assembly locating requirements at this accurate compensation stage, there is a difference on data collection compared to the experiment in Section 4.5. The data acquisition interval is not uniformly distributed across the entire motion stroke, considering the actual positioning scene/requirements. The data samples have a high density near the locating area. Considering the length of this chapter, not all the measured data samples are listed here. For example, only the measured training data set for x motion axis is shown partially in Table 4-9.

Table 4-9　The measured training data set for x motion axis

No.	x input /mm	x output /mm	y output /mm	z output /mm	No.	x input /mm	x output /mm	y output /mm	z output /mm
...	......	......	......	......	26#	166	166.0703	0.0208	−0.0924
15#	110	110.0443	0.0171	−0.0853	27#	168	168.0732	0.0177	−0.0909
16#	120	120.0502	0.0164	−0.0844	28#	169	169.0672	0.0282	−0.0927
17#	122	122.051	0.0174	−0.09	29#	170	170.0715	0.0205	−0.0922
18#	123	123.0482	0.0218	−0.0891	30#	171	171.0677	0.0265	−0.0907
19#	124	124.0504	0.018	−0.0852	31#	172	172.0715	0.0201	−0.0912
20#	125	125.0528	0.0165	−0.0874	32#	174	174.0727	0.0206	−0.0939
21#	126	126.0492	0.0242	−0.0875	33#	178	178.0714	0.0267	−0.0926
22#	130	130.0522	0.0245	−0.0895	34#	179	179.0754	0.0199	−0.0904
23#	140	140.0581	0.0222	−0.0881	...	......	......	......	......

Following the same method in section 4.5, the number of neuron nodes in hidden layer is set to 56, then the parameter identification results of this new SLFN model for error generator is shown as follows.

$$b = [0.354 \quad 0.961 \quad \cdots \quad 0.136]_{1\times168}^{\mathrm{T}} \tag{4-27}$$

$$w = \begin{bmatrix} -0.731 & 0.378 & 0.254 \\ 0.712 & 0.482 & 0.496 \\ \vdots & \vdots & \vdots \\ -0.673 & -0.349 & 0.345 \end{bmatrix}_{168\times3}^{\mathrm{T}} \tag{4-28}$$

$$v = \begin{bmatrix} 97492.749 & -13792.512 & 67422.924 \\ -87432.544 & 13257.856 & 14675.242 \\ \vdots & \vdots & \vdots \\ 94854.312 & 53437.825 & -31548.958 \end{bmatrix}_{168\times3}^{\mathrm{T}} \tag{4-29}$$

With the verification through calculations, for the fitting ability of the error generator model, its mean square deviation is 0.007mm. With the analysis of the testing samples, it is found that the mean square deviation of the prediction results is 0.002mm and the maximum mean deviation is 0.016mm. And the above parameters for the new SLFN prediction model, i.e., error generator, can be used in this accurate compensation stage.

4.6.3 Experiment Verification for the Accurate Compensation with Semi-mechanism Model

Under the instruction of Figure 4-15, in order to verify the effectiveness of the proposed semi-mechanism model, eight groups of positioning data that collected in practical assembly site are analyzed. With the measurement operations in CATIA environment and at the practical assembly site, the corresponding data is recorded and calculated, and the final positioning error based on accurate compensation with semi-mechanism analysis, is shown in Table 4-10.

With the analysis of the locating data in Table 4-10, after the accurate compensation with semi-mechanism motion model, the maximum positioning error is 0.042mm, the minimum value is 0.029mm, and the average error is 0.0356mm($\approx$0.036mm). Compared to the average error at the pre-compensation stage, of 0.0485mm, there is an enhancement ratio of 25.77%. This promotion indicates the positioning error is more accurate than the error obtained at the pre-compensation stage. In addition, the calculated error is closer to the measured practical value. In conclusion, the benefits and effectiveness that mentioned for semi-mechanism model are verified, and this model can be applied to guarantee the positioning accuracy for flexible assembly tooling.

Table 4-10　Accurate compensation with semi-mechanism analysis

No.	Theoretical position/mm	Initial input /mm	Output position (mechanism model)/mm	Output positioning error (data model) /mm	Modified input/mm	Final measured position/mm	Final positioning error/mm
1	3923.447, 1814.802, 1070.619	81.033, 93.764, 12.053	3923.601, 1815.156, 1070.847	0.165, −0.294, 0.211	80.868, 94.058, 12.264	3923.430, 1814.788, 1070.599	−0.017,−0.014, −0.020 ΔP=0.030
2	2874.139, 2020.168, 1254.671	87.015, 47.946, 16.033	2874.326, 2020.438, 1254.884	0.062, −0.107, 0.198	87.077, 48.053, 16.231	2874.132, 2020.144, 1254.685	−0.007,−0.024, 0.014 ΔP=0.029
3	3021.543, 1899.697, 1145.173	124.975, 28.348, 56.310	3021.789, 1900.000, 1145.355	0.113, 0.084, −0.117	125.221, 28.651, 56.492	3021.552, 1899.670, 1145.189	0.009,−0.027, 0.016 ΔP=0.033
4	3124.551, 2145.722, 1179.811	124.381, 78.473, 56.625	3124.764, 2145.974, 1180.000	0.016, −0.022, 0.007	124.397, 78.451, 56.632	3124.569, 2145.703, 1179.833	0.018,−0.019, 0.022 ΔP=0.034
5	3089.347, 2010.684, 1207.897	170.658, 98.624, 88.351	3089.375, 2010.678, 1208.348	−0.220, 0.015, −0.048	170.438, 98.639, 88.303	3089.335, 2010.710, 1207.929	−0.008,0.026, 0.032 ΔP=0.042
6	3127.018, 2131.427, 1295.014	179.994, 112.000, 124.972	3127.366, 2132.156, 1294.913	0.233, 0.357, −0.335	180.227, 112.357, 124.637	3127.052, 2131.423, 1295.001	0.034,−0.004, −0.013 ΔP=0.037
7	3296.754, 2274.877, 1056.683	228.957, 185.368, 99.324	3297.156, 2274.629, 1056.918	−0.532, 0.370, 0.344	229.425, 185.738, 99.668	3296.727, 2274.860, 1056.660	−0.027,−0.017, −0.023 ΔP=0.039
8	3394.577, 2302.241, 1208.899	265.393, 201.336, 135.000	3394.321, 2302.214, 1208.521	0.348, 0.021, 0.588	265.741, 201.357, 134.588	3394.570, 2302.208, 1208.880	0.007, 0.033, −0.019 ΔP=0.039

4.6.4　Discussion

For the comprehensive combination of mechanism model and data model, its characteristics and beneficial results can be gained.

Firstly, this solution can reduce the dependence of data items, when establishing SLFN prediction model. Correspondingly, the monitoring costs can be reduced. For example, with the consideration of the product features that needed to be assembled, more measurement operations are carried out in the nearby area of the location feature. Undoubtedly, the prediction results would be more accurate.

Secondly, combining the higher predictive ability of data model, with data processing and training operations, then the predicted positioning error based on original moving value of each motion axis, can be taken as an error generator. It can overcome the disadvantages of the five assumptions and the not very accurate calculation results. With the modified moving value and the error transfer mechanism model, the accurate compensation for the positioning error can be accomplished. The data model also becomes the basis of decision-making and serves the optimization of mechanism model. Then the simplified mechanism model can be used to represent the deterministic part of the accurate model, namely the cumulative transfer process of geometric errors[42]. This solution makes the relationship between causal factors and output result more comprehensive, and the overall effect is more consistent with the actual situation, including a higher precision value and a more accurate input/output relationship. The advantages of good interpretability for mechanism model can be gained for the comprehensive model.

Thirdly, with the accurate adjustment to mechanism model with the help of data model, the installation method of real-time measurement with the assistance of laser tracker, can be abandoned in practical flexible assembly site, and the effect of intelligent control can be achieved and verified.

4.7 Summary

In this chapter, the positioning error guarantee method with two-stage compensation strategy for aircraft flexible assembly tooling is studied, and the positioning error can be propagated accurately comparing to the measured results in practical assembly site. The main conclusions are shown as below.

① Based on the structure and working conditions of locating unit, an assembly measuring field is set up. The actual manufacturing error status of the comprised parts, the working conditions of the tooling locating unit, the various time-varying state parameters, and assembly errors of the product in practical assembly process, can be acquired.

② The error pre-compensation stage based on the error mechanism model is proposed, including the transfer and accumulation of actual geometric error status and physical deformation analysis. For the locating unit, based on the structure characteristics of the comprised parts and motion modules, considering the loading factors in the assembly process, the positioning error model for each motion axis and the entire locating space is established. With the analysis of the positioning accuracy for the four groups' key assembly characteristics, the variation range of the compensated error is within 0.05mm, which can basically meet the assembly requirements of aircraft parts in practical engineering.

③ The motion prediction model is proposed, considering the actual assembly conditions and the measured training/testing dataset of the positioning error. For the locating unit, the nonlinear and strongly coupled relationship between the input moving values and the output positioning errors is clarified, with ELM parameter identification in SLFN motion model. Instead of the error mechanism model, with the analysis of 480 groups of data samples, the propagation results of the SLFN prediction model are in accordance with the measurement errors in practical assembly site. For the end locators, SLFN model have an excellent fitting and prediction ability.

④ Combining the advantages of the error mechanism modeling and measurement data modeling, a semi-mechanism model is proposed for the accurate compensation of positioning error. Considering the influence of time-varying state parameters on the positioning accuracy of the tooling, a semi-mechanism model based on the measurement data is established, serving as an error generator. Where the mechanism model is taken as the basis, after the positioning error is predicted with SLFN model, compensated to the given input motion value, and the modified motion value is substituted in the mechanism model. The calculated positioning error is closer to the measured practical situation, with an enhancement ratio of 25.77% comparing to the pre-compensation stage. By adopting this accurate modification and compensation solution, a higher assembly efficiency and a convenient application effect would be gained.

⑤ Although the models of two compensation stages can be put into application to guarantee the positioning error for flexible assembly tooling, there are still problems remain. Firstly, only the effectiveness and accuracy of the positioning motion model are the main topics of this chapter. However, the computational efficiency should be enhanced due to the difficulties in disparate data acquisition and processing at practical assembly site[43]. Secondly, with the idea and solution for the accurate motion model, how to expressing the algorithm according to a software program with a certain modeling language, is a key step for integrating MES (manufacturing execution system). Thirdly, for the assembly environment and assembly process with a high dynamic change state, it's hard to verify the accuracy of the resulting models[44]. And the efficient and intelligent parameter identification of the motion model still needs further study.

References

[1] Moss S, Vezzetti E. Resistance spot welding process simulation for variational analysis on compliant assemblies[J]. Journal of Manufacturing Systems, 2015, 37(10): 44-71.

[2] Arista R, Falgarone H . Flexible best fit assembly of large aircraft components. Airbus A350 XWB case study[C]. Seville: Springer International Publishing, 2017.

[3] Tsutsumi D, Gyulai D, Kovács A, et al. Joint optimization of product tolerance design, process plan, and production plan in high-precision multi-product assembly[J]. Journal of Manufacturing Systems, 2020, 54(1): 336-347.

[4] Manohar K, Hogan T, Buttrick J, et al. Predicting shim gaps in aircraft assembly with machine learning and sparse sensing[J]. Journal of Manufacturing Systems, 2018, 48(C): 87-95.

[5] Ola, A. Quality modeling case study at GKN Aerospace Sweden[D]. Gothenburg: Chalmers

University of Technology, 2015.

[6] Cecil J, Mayer R, Hari U. An integrated methodology for fixture design[J]. Journal of Intelligent Manufacturing, 1996, 7(2): 95-106.

[7] Bakker O, Popov A, Ratchev S. Variation analysis of automated wing box assembly[J]. Procedia CIRP. 2017, 63: 406-411.

[8] Camelio J, Hu J, Ceglarek D. Impact of fixture design on sheet metal assembly variation[J]. Journal of Manufacturing Systems, 2004, 23(3): 182-193.

[9] Guo F, Zou F, Liu J, et al. Working mode in aircraft manufacturing based on digital coordination model[J]. The International Journal of Advanced Manufacturing Technology, 2018, 98: 1547-1571.

[10] Bullen G. Automated/mechanized drilling and counter sinking of airframes[M]. Warrendale: SAE International Press, 2013.

[11] Li X, Dang X, Xie B, et al. Flexible tooling design technology for aircraft fuselage panel component pre-assembly[J]. Assembly Autom 2015; 35(2): 166-171.

[12] Tadic B, Bogdanovic B, Jeremic B, et al. Locating and clamping of complex geometry workpieces with skewed holes in multiple-constraint conditions[J]. Assembly Automation, 2013, 33(4): 386-400.

[13] Erdem I, Helgosson P, Kihlman H. Development of automated flexible tooling as enabler in wing box assembly[J]. Procedia CIRP, 2016, 44: 233-238.

[14] Keller C, Putz M. Force-controlled adjustment of car body fixtures-verification and performance of the new approach[J]. Procedia CIRP, 2016, 44: 359-364.

[15] Fan W, Zheng L, Wang Y. An automated reconfigurable flexible fixture for aerospace pipeline assembly before welding[J]. The International Journal of Advanced Manufacturing Technology, 2018, 97(9-12): 3791-3811.

[16] Olabanji O, Mpofu K, Battaia O. Design, simulation and experimental investigation of a novel reconfigurable assembly fixture for press brakes[J]. The International Journal of Advanced Manufacturing Technology, 2016, 82(1-4): 663-679.

[17] Jamshidi J, Maropoulos P. Methodology for high accuracy installation of sustainable jigs and fixtures[M]. Advances in Sustainable Manufacturing. Berlin, Germany: Springer-Verlag, 2011: 149-155.

[18] Ramesh R, Mannan M, Poo A. Error compensation in machine tools-a review: Part I: geometric, cutting-force induced and fixture-dependent errors[J]. International Journal of Machine Tools & Manufacture, 2000, 40(9): 1235-1256.

[19] Tian W, Gao W, Zhang D, et al. A general approach for error modeling of machine tools[J]. International Journal of Machine Tools & Manufacture, 2014, 79(4): 17-23.

[20] Lin C, Ren Y, Ji J, et al. The bond graph method for analysis of the micro-motion characteristics of a micro gripper[J]. Strojniski Vestnik-Journal of Mechanical Engineering, 2016, 62(9): 494-502.

[21] Zhao D, Bi Y, Ke Y. Kinematic modeling and base frame calibration of a dual-machine-based drilling and riveting system for aircraft panel assembly[J]. The International Journal of Advanced Manufacturing Technology, 2018, 94(5-8): 1873-1884.

[22] Deng Y, Jin X, Zhang Z. A macro-micro compensation method for straightness motion error

and positioning error of an improved linear stage[J].The International Journal of Advanced Manufacturing Technology, 2015, 80(9-12): 1799-1806.

[23] Patel A, Ehmann K. Volumetric error analysis of a stewart platform-based machine tool[J]. Annals ofthe CIRP, 1997, 46(1): 287-290.

[24] Grandjean J, Ledoux Y, Samper S. On the role of form defects in assemblies subject to local deformations and mechanical loads[J]. The International Journal of Advanced Manufacturing Technology, 2013, 65(9-12): 1769-1778.

[25] Schleich B, Wartzack S. How to determine the influence of geometric deviations on elastic deformations and the structural performance?[J]. Proceedings of the Institution of Mechanical Engineers, Part B: Journal of Engineering Manufacture, 2013, 227(5): 754-764.

[26] Lustig R, Hochmuth R, Meerkamm H. Enhancement in coupling tolerance analysis and elastic deformations on the example of a serial linear support system [C]. Proceedings of the 15th International Conference on Engineering Design, Melbourne, Australia, 2005.

[27] Hao X, Li Y, Cheng Y, et al. A time-varying geometry modeling method for parts with deformation during machining process[J]. Journal of Manufacturing Systems, 2020, 55(4): 15-29.

[28] Aguado S, Santolaria J, Samper D, et al. Improving a real milling machine accuracy through an indirect measurement of its geometric errors[J]. Journal of Manufacturing Systems, 2016, 40(1): 26-36.

[29] Wang J, Guo J, Zhang G, et al. The technical method of geometric error measurement for multi-axis NC machine tool by laser tracker[J]. Measurement Science and Technology, 2012, 23(4): 045003.

[30] Nubiola A, Bonev I A. Absolute calibration of an ABB IRB 1600 robot using a laser tracker[J]. Robotics and Computer-Integrated Manufacturing, 2013, 29(1): 236-245.

[31] He Z, Fu J, Zhang L, et al. A new error measurement method to identity six error parameters of a rotational axis of a machine tool[J]. International Journal of Machine Tools & Manufacture, 2015, 88(1): 1-8.

[32] Vinod P, Reddy T, Sajin S, et al. Real-time positioning error compensation for a turning machine using neural network[J]. Procedia Materials Science, 2014, 5: 2293-2300.

[33] Nguyen H, Zhou J, Kang H. A calibration method for enhancing robot accuracy through integration of an extended Kalman filter algorithm and an artificial neural network[J]. Neurocomputing, 2015, 151(3): 996-1005.

[34] Maropoulos P, Muelaner J, Summers M, et al. A new paradigm in large-scale assembly— research priorities in measurement assisted assembly[J]. The International Journal of Advanced Manufacturing Technology, 2014, 70(1-4): 621-633.

[35] Goos J, Louarroudi E, Pintelon R. Generalizing periodically time-varying measurements with a parameter-varying input-output model[C]. 2015 IEEE Instrumentation & Measurement Technology Conference Proceedings, 2015.

[36] Wen Y, Yue X, Hunt J, et al. Feasibility analysis of composite fuselage shape control via finite element analysis[J]. Journal of Manufacturing Systems, 2018, 46(1): 272-281.

[37] Guo F, Wang Z, Liu J, et al. Locating method and motion stroke design of flexible assembly tooling for multiple aircraft components[J]. The International Journal of Advanced Manufac-

turing Technology, 2020, 107(1-2): 549-571.

[38] Hu Y, Wang Z, Ren C, et al. Locating datum modeling for minimum normal error in aircraft digital assembly system[C]. Dordrecht: Atlantis Press, 2015.

[39] Baldi P, Sadowski P, Lu Z. Learning in the machine: The symmetries of the deep learning channel[J]. Neural Networks, 2017, 95(11): 110-133.

[40] Michael S, Ronay A, Thomas H. A survey of the advancing use and development of machine learning in smart manufacturing[J]. Journal of Manufacturing Systems, 2018, 48(C): 170-179.

[41] Mantoro T , Olowolayemo A , Olatunji S O , et al. Extreme learning machine for user location prediction in mobile environment[J]. International Journal of Pervasive Computing and Communications, 2011, 7(2): 162-180.

[42] Florian S, Verena H, Florian O, et al. Tolerance analysis of compliant, feature-based sheet metal structures for fixtureless assembly[J]. Journal of Manufacturing Systems, 2018, 49(10): 25-35.

[43] Tao F, Qi Q. Make more digital twins[J]. Nature, 2019, 573(7775): 490-491.

[44] Qi Q, Tao F, Hu T, et al. Enabling technologies and tools for digital twin[J]. Journal of Manufacturing Systems, 2021, 58: 3-21.

Chapter 5
Assembly Error Propagation Modeling and Coordination Error Chain Construction for Aeronautical Structure

5.1 Introduction and Related Work

Assembly is regarded as the process of putting or jointing parts together to form the final product according to their relative positions. It is of great importance in product life cycle, because it is intensively related to the development time, cost, and dimensional quality of the final product[1]. As manufacturing aeronautics products, a working mode based on technologies of digital tooling (DT)[2], guageless tooling (GT)[3], model based definition (MBD)[4], measurement assisted assembly (MAA)[5], digital twin[6], smart factory[7], and others, has been widely used in practical applications. Correspondingly, to sheet metal parts or those with streamline profile, CNC machining and digital measurement methods are used in the manufacturing process and their relative forming tooling process; in assembly tooling, special measuring equipment, such as laser tracker and grating sensor, is used in its installing and checking process; and in products' assembly work, numerous digital devices are used in the detailed locating, drilling, jointing, and detecting process. To sum up, the above actions can ① improve the accuracy of each error links in assembly process, ② simplify the dimension and shape transfer process, and ③ reduce the number of manufacturing error links for assembly procedure, which would enhance assembly accuracy for practical aeronautical productions. As a matter of fact, although the assembly accuracy of the skin profile and hinge joints for certain component can be guaranteed, the consistency of accumulated assembly errors for different components, known as coordination error, such as ① the gap and flush between different profiles, or ② the coaxiality between different hinge joints, is always difficult to satisfy the design requirements. It is mentioned that

Digital Assembly Coordination and Quality Controlling Technology
for Aeronautical Thin-walled Structures

coordination error has a direct influence on the property of stealthy and reliability for modern aircrafts[8]. If the coordination error is out of control, intensive manual repair works on coordination regions would be carried out, and the repair workload would be much heavier for parts or components with titanium alloy or composite material. The scrapping rate of certain assemblies can reach up to about 50% as dealing with the inconsistency of accumulated assembly errors[9].

With the inconsistency phenomenon that mentioned above, it can be explained that the transfer mechanisms from assembly error and coordination error are different. Assembly error is the organic combination of the errors from parts, fixtures, locating and jointing operations. In practical application, if its cumulative value falls in a reasonable variation range, then the error can meet the design requirement[10]. However, the fact is that a qualified assembly error may not result in a satisfactory coordination error. And there is a close relationship between them, as coordination error is reflected by the consistency of accumulated assembly errors between different components, in both the accumulating direction and value size. However, limited attention has been paid on this phenomenon. Many factors, such as ① transformation of assembly stations, state, and datum, ② accumulation of large amount of error links (about twice as many as assembly error), and ③ influence among all kinds of error items should be taken into account as constructing coordination dimension chain to improve manufacturing process. But the fact is that the requirements of coordination error is always stricter than those for assembly error items. If there is not an accurate accumulated assembly error, coordination error can be approximated as the sum of the absolute values of assembly error ranges. This would make the numerical value of coordination error larger than assembly error, and it also prevent accurate advanced calculation and evaluation. As a result, intensive manual repair works are often needed in practical assembly sites to guarantee the coordination accuracy[11]. Based on the above analysis, more accurately modeling and predicting assembly errors in both the accumulation size and direction, is an efficient method to control coordination accuracy. Aiming at evaluating coordination error, effective efforts, such as studying the influence relationships of different error items analyzing and error propagation modeling, have been done in pertinent literature.

To model the basic error items, Li[12] analyzed the tolerance using small displacement torsor (SDT) with GD&T. Guo[13] described a unified error model for various types of geometric tolerance, obtaining the distribution of the deviations for the control points along different directions. Then, according to ANSI or ISO Standards, Franciosa[14] automatically calculated the variational parameters for planar or cylindrical features based on a given set of tolerance specifications. It is mentioned that, only the value of each dimensional error is not enough to calculate the final assembly variation.

Since assembly process is continuous, error items will couple with each other and accumulate as the process progresses. To model assembly error, Liu[15] proposed a state space model (SSM) for 3D variation propagation in multistage assembly processes, containing various types of variation sources. With stream-of-variation analysis (SOVA), Huang[10] established the mathematical models for 3D single-station and multi-station assemblies. In addition, based on differential motion vectors (DMV) with homogeneous transformations, Du[16] developed a generic framework for 3D variation propagation modeling in multistage turning processes. Although the above findings could provide impactful methods for assembly error modeling, they are mainly based on rigidity assumption, considering only rigid variations. For most aviation/aerospace parts and components with low rigidity, deformation is inevitable in each assembly process. As a result, assembly error propagation with the above methods cannot be modeled accurately and effectively.

Taking the factor of assembly deformation into account, Bakker[17] analyzed the flexible factors of parts in automatic assembly of a wing-box, and SOVA method was used to analysis assembly deviation, then the maximum repair value between upper and lower panels was derived. Das[18] developed an assembly transfer function (ATF) approach to model and simulate assembly process with compliant non-ideal parts. Franciosa[19] presented a finite element model (FEM)-based computing tool for statistically analyzing variations of flexible parts that occurring in assembly processes. To better evaluate the dimensional quality of sheet metal assembly with generic 3D free surface, with differential motion vector and homogeneous transformation matrix, Cai[20] proposed a rigid-compliant hybrid variation modeling methodology based on the method of influence coefficients (MIC). To improve computational efficiency for compliant variation, Camelio[21] predicted assembly variation of compliant parts with principle component analysis (PCA) and FEM analysis. Then, to reduce the modeling complexity, Lin[22] proposed a deviation propagation model (DPM) with FEM, condensing compliant parts into several substructures, however, DPM has a limited application range for complex assemblies with variable structures. In summary, with above mathematical models, the calculation results are more accurate compared to rigidity assumption. However, the relationships between different error items at different assembly stages are simplified to linear, making the nonlinear accumulating characteristics between assembly deviation and variation sources be not thoroughly analyzed.

In the nonlinear accumulation process for assembly/coordination error, Qu[23] established a variation propagation method with discrete nonlinear state-space model in multi-station assembly process for sheet metal products. Then, to explicitly describe the nonlinear behavior of physical interaction function in compliant components, Wang[24]

Digital Assembly Coordination and Quality Controlling Technology
for Aeronautical Thin-walled Structures

proposed an assembly deformation prediction model and a variation propagation model to calculate assembly error of aircraft panels, deriving the consecutive 3D deformation expressions. Considering the influences of manufacturing errors on product shape errors, Li[25] proposed a method based on primitive deformation patterns to build variation propagation model. To control assembly gaps for aircraft wing, Wang[26] proposed an optimal posture evaluation model with preliminary and refined alignments, and a comprehensive weighted minimization model with gap tolerance constraints was established for redistributing the gaps in multi-regions. In addition, to improve dimensional quality, with ① off-line error control-learning module for determining necessary adjustments based on virtual assembly models, and ② in-line implementation with a feed-forward control strategy based on the above learned adjustments, Xie[27] proposed a complete method for dimensional-related error compensation in the assembly process of compliant sheet metal parts. Summarizing the above research, the error modeling and deviation propagation mechanism for different kinds of products are presented, analyzing the nonlinear characteristic explicitly and efficiency. However, due to the characteristics of ① multi-station and multi-hierarchy for assembly process, ② multi-transformation for manufacturing datum, and ③ ubiquitous deformation errors for products and tooling, the precise assembly is extremely difficult to fulfill for aircraft. To guarantee the precise propagation of assembly error and to control the coordination errors among different components, substantial works remains. Firstly, the influence relationship among different error items, such as the coupling principle between assembly deformation and positioning error, needs to be further analyzed. Secondly, since the assembly error has a direct relationship with dimension error and positioning error, to improve the accuracy of calculation results, actual position of the key features is to be determined in a unified coordination system according to the practical assembly process and physical state. Thirdly, as the assembly process flowing down, in the transformation process for different kinds of coordination datum, such as ① machining datum, ② locating datum, ③ assembly datum, and ④ measuring datum, spatial position for the key features of coordination regions needed be modified according to the practical manufacturing situations.

In this chapter, the objective is to solve the error problems by modeling the assembly error propagation process and constructing the coordination dimension chain to control the consistency of accumulated assembly errors. By revealing the transfer mechanism of assembly error, a set of analytical methods for the construction of coordination error chain is developed, with the ultimate goal of realizing precising assembly for complex products, such as aircraft and automobile. In Sections 5.2 and 5.3, three basic error sources and their interaction relationships are modeled. Section 5.4 presents the error model in transformation process for coordination datum. In Section 5.5, error propagation process for a single assembly station and for multiple assembly stations are precisely modeled, to gain the coordination error chain between different assemblies, obtaining the final coordination error. Section 5.6 presents a practical engineering case to show the methodology's feasibility. Conclusions and further works are shown in the final section.

5.2 Basic Error Sources Modeling in Assembly Process

During the assembly process of complex products, there are mainly three kinds of basic error sources: manufacturing error of parts, positioning error of assembly tooling, and deformation error when assembling parts or components.

5.2.1 Part Manufacturing Error

In the forming or machining process for different kinds of sheet metal parts or machined parts, manufacturing errors are inevitable. The error ξ_p refers to the deviation of dimension or shape between actual and theoretical comprised geometric features for a manufactured part. Specifically, it is the dimensional and form error in its local coordinate system. According to the representation method of GD&T (geometric dimensioning & tolerancing) and SDT [12], by adding the feature dimension set S, manufacturing error can be expressed by a binary set of H, as shown in Equation (5-1). Where $d_i(i = x, y, z)$ stands for the transition error, and $\theta_i(i = x, y, z)$ represents the rotation error.

$$H = \{\tau, S\} = \{D, \Omega, S\} = \{[d_x \ d_y \ d_z], [\theta_x \ \theta_y \ \theta_z], S\} \tag{5-1}$$

Actually, the above key features have a high manufacturing precision requirement, and also have a close relationship with assembly error. As manufacturing an aircraft, ① design/process interface, ② regions on interchangeable parts or benchmark parts, ③ the same hinge joints or contour profiles that located by a serial assembly jigs, ④ contour profile with a complex surface, and ⑤ regions or parts that both have a manufacturing relationship with its forming die and a coordination accuracy relationship with its assembly tooling, are the key coordination regions. Correspondingly, geometric features, such as ① straight lines (beam axis, wing axis, bulkhead axis, and stringer axis, etc.), ② flat surface and cylindrical surfaces (butt joints and frame-type hinge joints, etc.), and ③ free-form surfaces (aerodynamic profile), etc. are the main feature classification. For different geometric features, S can either be a null set or a combination of many dimensions. For $\{\tau\}$, the comprised elements in the constant direction are zero.

The value of manufacturing error can be gained through certain types of measuring operations. According to the experience and statistical data in manufacturing, the above basic error items often obey a normal probability distribution, i.e., $\xi_p \sim N(\mu_p, \sigma_p)$, and are typically systematic in positive and negative directions when the accumulated assembly variation is analyzed and calculated[28]. It is mentioned that the deformation issue at assembly stage is considered in this chapter. However, to manufacturing deformation, its generating principle is not concerned here. Although the manufacturing deformation of part leads to assembly variations eventually[29], to simplify the modeling complexity of assembly error propagation, manufacturing deformation is considered only as an unchangeable geometrical error source, i.e., dimension, shape and form error items, when parts are assembled at the locating and jointing stage. As a result, the detailed influence of manufacturing deformation on assembly deformation is not taken into consideration during the assembly process in this chapter, and the part's reference also remains the same in the whole manufacturing process.

5.2.2 Positioning Error of Assembly Tooling

Positioning error of assembly tooling has a direct influence on assembly error and coordination error. For assembly tooling, the end locaters are often installed on its corresponding fixed pedestal after being adjusted to meet the desired accuracy requirement. Excluding the environment factors, the positioning error ξ_f of assembly tooling can also be regarded as a constant value during the whole assembly process. However, this type of error may refer to the position variation in locating holes or mating surfaces, which has a direct geometric coordination relationship with the products to be assembled. Similar to part manufacturing error, positioning error can also be taken as a normal probability variable according to the measured results from the practical installation site, which has a distribution of $\xi_f \sim N(\mu_f, \sigma_f)$. And the detailed error magnitude can also be represented by SDT.

5.2.3 Assembly Deformation Error

For complex parts and components with low rigidity and special aerodynamic shapes or structures, deformation is likely to occur in the assembly process. And it is caused by different kinds of assembly forces during assembly operation, such as clamping force, drilling force, gravity force, jointing force, and resilience force, as shown in Figure 5-1.

To ensure the final assembly accuracy and avoid the assembly rework, assembly deformation error ξ_d must be analyzed and controlled to support the entire propagation process of assembly/coordination errors. In the modeling process of assembly

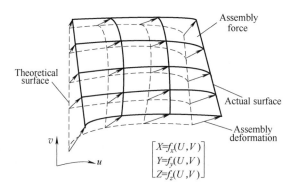

Figure 5-1 Assembly deformation error for freeform surface

deformation, with the construction of stiffness matrix for a given object, the relationship between assembly force and assembly deformation can be gained according to the finite element simulation method. Assuming that there are n action points associated with assembly force, if a unit force F is applied at the jth action point, the deformation response values U_j of all action points can be obtained. Similarly, if a unit force F is applied to all n action points respectively, the deformation response values can be gained by Equation (5-2). Where matrix $[C]_{n \times n}$ stands for the response between assembly force and corresponding assembly deformation, and is the inverse expression of stiffness matrix $[K]$.

$$\{U\} = \left\{ \begin{matrix} U_1 \\ U_2 \\ \cdots \\ U_n \end{matrix} \right\} = \begin{bmatrix} c_{11} & c_{12} & \cdots & c_{1n} \\ c_{21} & c_{22} & \cdots & c_{2n} \\ \vdots & \vdots & \ddots & \vdots \\ c_{n1} & c_{n2} & \cdots & c_{nn} \end{bmatrix} \left\{ \begin{matrix} F_1 \\ F_2 \\ \vdots \\ F_n \end{matrix} \right\} = [C]\{F\} = [K]^{-1}\{F\} \qquad (5\text{-}2)$$

Similarly, if a random force F', instead of F, is applied, then the deformation error U' at these action points can be expressed by Equation (5-3).

$$\{U'\} = [C]\{F'\} \qquad (5\text{-}3)$$

However, no matter the assembly deformation error ξ_d of different parts or components caused by any kinds of assembly force, the results can be also expressed according to Equation (5-1), i.e., $\xi_d = \{\xi_{dx}, \xi_{dy}, \xi_{dz}, \xi_{\theta x}, \xi_{\theta y}, \xi_{\theta z}\}^T$.

It is mentioned that, ① the stiffness matrix $[K]$ is a variable parameter as the assembly process flowing down, as the number of parts in a component increases dynamically; ② in order to take full advantage of assembly deformation information, such as assembly sequence planning, assembly error simulating, or assembly process optimizing, it is necessary to reconstruct the deformed geometric features (such as

Digital Assembly Coordination and Quality Controlling Technology
for Aeronautical Thin-walled Structures

freeform surface, cylinder surface, flat surface, and others) according to the deformation at certain key discrete points. Once the operation work at each assembly stage is finished, the model can be reconstructed and used for assembly analysis in the next assembly stage, to enhance the reliability of assembly error propagation results. An efficient solution for practical surface reconstruction is bicubic B-spline surface interpolation, and the detailed process is shown in Figure 5-2.

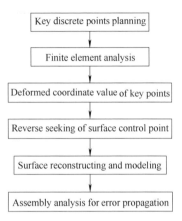

Figure 5-2   Surface reconstruction with assembly deformation

5.3   Interaction Relationship among Different Basic Error Items

As well known, the structures, topological shapes, and spatial relationships are complex for aircraft products. Due to ① a huge amount of relevant assembly tooling or fixtures, and ② a long manufacturing process, error items, such as dimension/shape/position errors, would interact and couple with each other as assembly process flowing down. As a result, the propagation process for assembly errors would demonstrate a nonlinear accumulating characteristic between assembly errors and variation sources. In the detailed locating—clamping—jointing—releasing assembly stage, manufacturing error of parts or components would be combined with positioning errors of assembly tooling to form a mating error δ_q, which is mainly caused by geometric errors. However, at different assembly stations/stages, or during different working procedures at a certain assembly stage, the mating error δ_q at previous assembly stage may be an input error for the following assembly deformation error ξ_d, and ξ_d at previous assembly stage may also be an input error for another mating error δ_q, making the propagation of assembly error complex to model. To analyze the assembly/coordination errors accurately, the mutual influence relationships among the basic errors are modeled in this section.

5.3.1 Mating Error

In the assembly locating process, the error of geometric features that distributed on different parts and assembly tooling would couple with each other to form a desired accuracy of products. As the existence of ① manufacturing error of parts, and ② positioning error of the corresponding assembly tooling, parts that located on assembly tooling would be translated or rotated in the assembly coordinate system, causing the key features to a posture change from the desired position. This change is called as mating error δ_q, which can be calculated by Equation (5-4). It is mentioned that δ_q can also be expressed by SDT and shown as a vector, i.e., $\delta_q = \{\delta_{qx}, \delta_{qy}, \delta_{qz}, \delta_{q\alpha}, \delta_{q\beta}, \delta_{q\gamma}\}^{\mathrm{T}}$. Where, Φ represents the constraint vector of the parts, and it is expressed in $\Phi = [0, \cdots, n_i^{\mathrm{T}}, \cdots, 0]^{\mathrm{T}} (i = 1, 2, \cdots, 6)$; $[S]$ stands for the error sensitivity matrix, indicating the accuracy changing response to the key features when a unit deviation is applied; J represents the Jacobi matrix, and $J_f = [J_1, \cdots, J_6]^{\mathrm{T}}$, each J_i can be calculated by Equation (5-5). With the analysis methods of Monte Carlo simulation and finite element simulation, the statistical properties of δ_q can be gained.

$$\delta_q = -J^{-1}\Phi(\xi_f - \xi_p) = [S](\xi_f - \xi_p) \tag{5-4}$$

$$J_i = \left[\frac{\delta\Phi_i}{\delta x_0} \quad \frac{\delta\Phi_i}{\delta y_0} \quad \frac{\delta\Phi_i}{\delta z_0} \quad \frac{\delta\Phi_i}{\delta\Psi} \quad \frac{\delta\Phi_i}{\delta\theta} \quad \frac{\delta\Phi_i}{\delta\Phi} \right] \tag{5-5}$$

5.3.2 Influence of Matting Error on Deformation Error

Locating is the first operation in the detailed assembly process, and it is also the fundamental to analyze the errors of the remaining operations. To analyze the influence relationship of mating errors on deformation errors, for the key features on the product, propagation process of the error items at different assembly stations or stages is analyzed firstly, as shown in Figure 5-3. Where, θ_i stands for the influence error on deformation error ξ_d that caused by matting error δ_q at previous assembly procedure; λ_i represents the final assembly error with the consideration of θ_i.

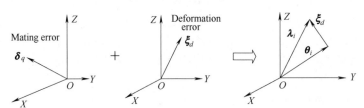

Figure 5-3 Influence of matting error on deformation error

And the above influence relationship can be modeled by Equation (5-6).

$$\lambda_i = \xi_d + \theta_i = \xi_d + M_1 \cdot \delta_q \tag{5-6}$$

From Equation (5-6), it can be known that the solution of θ_i can be translate to solving the error influence matrix M_1. Given the consideration of the u direction for key features of a certain part, the influence value at the ith key feature has a relationship with the mating errors of its two adjacent features, as shown in Figure 5-4. Where δ_{q2i}^{i-1} and δ_{q2i}^{i+1} stand for the mating errors of the $(i-1)$th and $(i+1)$th key features at the theoretical positions of L_{2i}^{i-1} and L_{2i}^{i+1} along the given direction u, and $L_{2i}^{i\prime}$ stands for the theoretical position of the ith key feature.

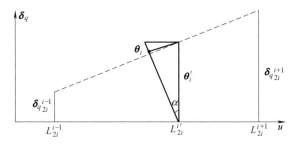

Figure 5-4   Projection of the ith key feature on u direction

By projecting the ith key feature on the reconstructed deformed model on to a line that connecting the $(i-1)$th and $(i+1)$th key features, with the consideration of geometric continuity principle, then the value of θ_i' that representing the projection of the ith key feature's mating error can be given by Equation (5-7). Where, α represents the projection angel between the practical position in reconstruction model and theoretical position of ith key feature with direction u.

$$\theta_i' = \frac{L_{2i}^{i\prime} - L_{2i}^{i-1}}{L_{2i}^{i+1} - L_{2i}^{i-1}} \cdot (\delta_{q2i}^{i+1} - \delta_{q2i}^{i-1}) + \delta_{q2i}^{i-1} \tag{5-7}$$

Taking $(L_{2i}^{i\prime} - L_{2i}^{i-1})/(L_{2i}^{i+1} - L_{2i}^{i-1})$ as L_2^i, and the value of the influence error θ_i for the ith key feature can be given by Equation (5-8).

$$\theta_i = \frac{L_2^i}{\cos \alpha} \cdot \delta_{q2i}^{i+1} - \frac{L_2^i - 1}{\cos \alpha} \cdot \delta_{q2i}^{i-1} - \delta_{q2i}^i \tag{5-8}$$

With the above analysis, the error influence matrix M_u can be expressed by Equation (5-9).

$$M_u = \begin{bmatrix} M_{11} \\ M_{12} \end{bmatrix} = \begin{bmatrix} L_2^i / \cos \alpha \\ (L_2^i - 1) / \cos \alpha \end{bmatrix} \tag{5-9}$$

It is mentioned that, for the key features at v direction for a certain part, the error influence matrix can also be solved with the above solution. Then, the final influence matrix can be expressed by Equation (5-10). Where, β represents the projection angel with direction v.

$$M_1 = \begin{bmatrix} M_u \\ M_v \end{bmatrix} = \begin{bmatrix} L_2^i / \cos\alpha & L_2^{i'} / \cos\beta \\ (L_2^i - 1) / \cos\alpha & (L_2^{i'} - 1) / \cos\beta \end{bmatrix} \qquad (5\text{-}10)$$

5.3.3 Influence of Deformation Error on Matting Error

To analyze the influence of deformation error on mating error at different assembly stations or stages, the propagation process of the ith key feature in the assembly process is expressed in Figure 5-5. Where, ρ_i' stands for the influence error on mating error δ_q' that caused by deformation error ξ_d' at previous assembly stage; λ_i' represents mating error of the current assembly stage; and ϑ_i' represents the final assembly error with consideration of ρ_i'.

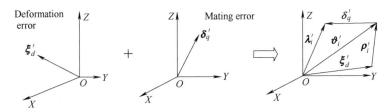

Figure 5-5   Influence of deformation error on matting error

And the above influence relationship can be modeled by Equation (5-11).

$$\lambda_i' = \delta_q' + \vartheta_i' = \delta_q' + \xi_d' + \rho_i' \qquad (5\text{-}11)$$

To solve the error ρ_i', an analysis work for the decomposition of above influence error in to three axial directions of a 3D Cartesian coordinate system should be done firstly. The deformation error ξ_d' at the previous assembly stage would affect the present mating error λ_i' in both the direction of translation and rotation relative to the coordinate axis. Then the error ρ_i can be given in Equation (5-12). The solution of influence relationship between deformation error and mating error can be modeled by calculating the influence matrix M_2.

$$\rho_i = M_2 \cdot \xi_d' = M_{2T} \cdot M_{2R} \cdot \xi_d' \qquad (5\text{-}12)$$

Where M_{2T} is the error transformation matrix with the translational direction, M_{2R} is the matrix with the rotational direction, as given by Equation (5-13) and Equation (5-14), respectively.

Digital Assembly Coordination and Quality Controlling Technology
for Aeronautical Thin-walled Structures

$$M_{2T} = \begin{bmatrix} 1 & 0 & 0 & x \\ 0 & 1 & 0 & y \\ 0 & 0 & 1 & z \\ 0 & 0 & 0 & 1 \end{bmatrix} \qquad (5\text{-}13)$$

$$M_{2R} = M_{2Rx} M_{2Ry} M_{2Rz}$$

$$= \begin{bmatrix} 1 & 0 & 0 & 0 \\ 0 & \cos x & -\sin x & 0 \\ 0 & \sin x & \cos x & 0 \\ 0 & 0 & 0 & 1 \end{bmatrix} \begin{bmatrix} \cos y & 0 & \sin y & 0 \\ 0 & 1 & 0 & 0 \\ -\sin y & 0 & \cos y & 0 \\ 0 & 0 & 0 & 1 \end{bmatrix} \begin{bmatrix} \cos z & -\sin z & 0 & 0 \\ \sin z & \cos z & 0 & 0 \\ 0 & 0 & 1 & 0 \\ 0 & 0 & 0 & 1 \end{bmatrix}$$

$$(5\text{-}14)$$

By introducing the error influence matrix M_2, then ρ_i can be converted from the current coordinate system to the coordinate system of the specific product object for further assembly error analysis.

5.4 Error Modeling for Transformation Process of Coordination Datum

There are four types of coordinate systems (CS) in the design and manufacturing process for complex products such as aircraft: CS of the whole aircraft, auxiliary CS (such as CS for fuselage or wing), local CS for parts, and CS for assembly tooling or jigs. These CSs are the basis of determining the position of parts and components at different manufacturing and assembly process. In this section, the goal is to gain the error in coordination datum transforming process by modeling ① practical establishment error of assembly coordinate system and actual spatial position of parts or components, and ② relocation errors and modification for key product features in actual assembly site under practical assembly state.

5.4.1 Definition-coordination Datum Transformation

Considering the procedures and characteristics of assembly process, coordination datum transformation is defined as the change of reference coordinate system, assembly datum, locating datum, or machining datum for certain assembly units at different assembly stations or hierarchies. It is mentioned that, if there is an original error for the above datum, the variation of relevant key features would have a significant amplification effect as assembly operations flowing down.

5.4.2 Establishment of Assembly Tooling CS and Spatial Position Modeling

In digital manufacturing environment, installation operations for assembly tooling are carried out using its own actual CS. To establish tooling's CS, firstly, the mathematical model of components is put into a specific measurement system, then, the theory coordinate data of optical tooling ball (OTP) points that distributed on assembly platform is gained. Three target measuring points, TB1, TB2, and TB3, are taken as the datum to establish the Cartesian coordinate system of the tooling. It is mentioned that the three tooling ball (TB) points, located at the corners of tooling's fundamental surface, have been positioned accurately at the beginning of installation operations. In practical assembly site, to establish the coordinate system accurately, some enhanced reference system (ERS) points are also often used. The establishment error of assembly tooling CS refers to the deviation between actual posture of tooling CS and the theory posture, and positioning error ξ_f of ending locators has a close relationship with the precision of actual tooling CS, as shown in Figure 5-6 and Figure 5-7. Where, $OXYZ$ stands for theory CS, and $O'X'Y'Z'$ represents actual tooling CS. It is mentioned that to ensure the coordination relationship among different assemblies, an accurate transformation relationship between $OXYZ$ and the product design CS should be gained firstly.

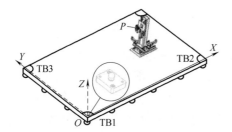

Figure 5-6   Coordinate system of the assembly tooling

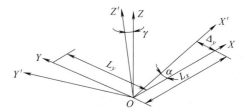

Figure 5-7   Axes errors in the tooling coordinate system

Regarding the practical establishment error of assembly CS, the actual rotation angle γ for Z-axis has a correlation with the coordinate values of the above three TB points at Z direction, and γ can be determined by the normal vector of actual XOY plane. Assuming that the coordinate of the first TB point is $(0,0,z_1)$, the coordinates of $(L_x,0,z_2)$, $(0,L_y,z_3)$ are for the other two TB points, then the normal vector $\boldsymbol{n}$ of the practical XOY plane can be gained by Equation (5-15).

$$\boldsymbol{n}=[L_y(z_1-z_2),L_x(z_1-z_3),L_xL_y] \tag{5-15}$$

Then the actual rotation angle α and β around X-axis and Y-axis can be represented with Equation (5-16), where, Δx and Δy represent position errors of the measuring points along the directions of X and Y axes, respectively, L_x and L_y represent the distance between the measuring points along the above two axes.

$$\begin{cases} \Delta\alpha \approx \Delta_y / L_x \\ \Delta\beta \approx \Delta_x / L_y \end{cases} \tag{5-16}$$

It is mentioned that α and β can be taken as the same value because of the use of axis alignment method in the measuring system, where the Y-axis is determined automatically by the pre-determined X-axis. However, the mean value of calculation results in Equation (5-15) and (5-16) can be used to analyze the practical establishment error of CS. According to Equation (5-16), for the assembly tooling with huge dimension and volume, to ensure a higher positioning accuracy and a lower establishment error of the tooling CS, the distance between TB points should be designed to be as long as possible.

Without considering the practical error of different CSs, to get the accurate spatial position of parts, the part CS should be rotated firstly to keep its directions in alignment with the assembly tooling CS, and then moved to the position of the tooling CS. The above whole transformation process can be expressed by a translation matrix T and a rotation matrix R, as shown by Equation (5-17).

$$y' = Ry + T \tag{5-17}$$

With the analysis based on Equation (5-15) and (5-16), it is known that the practical deviation of the homogeneous coordinate transition matrix can be calculated by Equation (5-18), according to the rotation angle variation for actual CS.

$$R' = R + \Delta R = [\alpha + \Delta\alpha, \beta + \Delta\beta, \gamma + \Delta\gamma]^{\mathrm{T}} \tag{5-18}$$

Due to the existence of assembly CS error, the actual spatial position of key features in practical assembly work would have a deviation from their theory position that defined in the theoretical MBD model, and the actual spatial position can be modeled by Equation (5-19).

$$y'' = R'y + T' \tag{5-19}$$

By analyzing the transformation relationships among different assembly CSs and combining them with the transformation relationships among product design CSs, the actual spatial position of key features can be gained in a unified CS. This would facilitate the analysis assembly error propagating process. Then the actual spatial position deviation that caused by the transformation error between the theoretical and actual assembly CSs can be calculated by Equation (5-20).

$$T_{i,j} = \Delta y = y'' - y' \tag{5-20}$$

It is mentioned that to decrease the transformation error among different assembly CSs, the positions of TB points should be detected at certain regular intervals.

5.4.3 Relocation Error Modeling and Modification of Key Product Features

To certain given parts or components at different assembly hierarchies, when the manufacturing datum doesn't stay the same as the assembly datum, or the locating datum changes at different assembly stations, a phenomenon of relocation transformation occurs for key product features. This misalignment would bring a spatial position error for relevant parts or components, i.e., the relocation error, donated as C_{ij}.

As shown in Figure 5-8, the first subassembly is comprised by two parts at the kth assembly station, namely part 1 and part 2, respectively. Combining with part 2, part 1 is taken as datum reference to finish assembly process, and the assembly features are F_k and F_l on part 1. Contrasting to part 1, the mating features are F_k' and F_l' on part 2. However, for part 1, the machining datum features may be F_i and F_j. Without considering the preexisting mating error of the key features on part 2, this datum misalignment would bring an extra error item for part 2, and its actual spatial position is shown by the dotted lines.

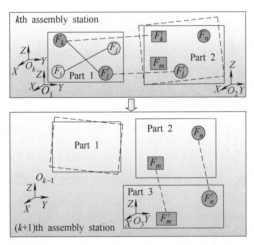

Figure 5-8 Generating principle of relocation error (See the color illustration)

In the following assembly process at the $(k+1)$th station, the above subassembly at the kth assembly station acts as a special part for the current assembly work, where another part, i.e., part 3, is to be assembled. However, part 1 is not the datum reference anymore at this assembly station, instead, part 2 is taken as the datum part here. In this relocation process, part 2 would deviate from its nominal position, and the assembly error at the previous assembly station would be reset. To finish the locating work for the above subassembly, different geometric mating features F_m and F_n on part 2 (not F_k' and F_l' anymore) would couple with F_m' and F_n' that distributed on part 3, respectively. Due to the existence of another misalignment for the locating datum between the above two assembly stations, an extra spatial position error, contrasting with the first subassembly, is brought out.

At the kth assembly station, relocation error is caused by the changes of datum geometric features. Assuming that the machining datum features F_i and F_j have spatial position errors of $(\Delta x_i, \Delta y_i, \Delta z_i)$ and $(\Delta x_j, \Delta y_j, \Delta z_j)$ under their own part CS, $O_1 X_1 Y_1 Z_1$, respectively. And the assembly datum features F_k and F_l have errors of $(\Delta x_k, \Delta y_k, \Delta z_k)$ and $(\Delta x_l, \Delta y_l, \Delta z_l)$ in $O_1 X_1 Y_1 Z_1$, respectively. Let L_{ij} represents the deviation distance of $\sqrt{(\Delta x_j - \Delta x_i)^2 + (\Delta y_j - \Delta y_i)^2 + (\Delta z_j - \Delta z_i)^2}$, and L_{kl} stands for $\sqrt{(\Delta x_l - \Delta x_k)^2 + (\Delta y_l - \Delta y_k)^2 + (\Delta z_l - \Delta z_k)^2}$. Then according to the computing rule of vectors, contrasting with three theoretical directions of part CS axes, the actual rotation deviations, namely angle α, β, and γ of part 1 that caused by datum transformation, can be calculated by Equation (5-21).

$$\begin{cases} \alpha \approx (\Delta x_j - \Delta x_i)/L_{ij} + (\Delta x_l - \Delta x_k)/L_{kl} \\ \beta \approx (\Delta y_j - \Delta y_i)/L_{ij} + (\Delta y_l - \Delta y_k)/L_{kl} \\ \gamma \approx (\Delta z_j - \Delta z_i)/L_{ij} + (\Delta z_l - \Delta z_k)/L_{kl} \end{cases} \quad (5\text{-}21)$$

Given the key feature m_1 that distributed on part 1 has a position coordinate of (x_{m1}, y_{m1}, z_{m1}) under its own manufacturing CS, $O_1 X_1 Y_1 Z_1$, and the position having a variation of $(\Delta x_{m1}, \Delta y_{m1}, \Delta z_{m1})$, then the position error for m_1 that caused by datum misalignment can be gained by Equation (5-22).

$$\Delta \boldsymbol{m}' = \boldsymbol{M}_{3\times3} \Delta \boldsymbol{m}_1 = \begin{bmatrix} 1 & 0 & 0 \\ 0 & \cos\alpha & -\sin\alpha \\ 0 & \sin\alpha & \cos\alpha \end{bmatrix} \begin{bmatrix} \cos\beta & 0 & \sin\beta \\ 0 & 1 & 0 \\ -\sin\beta & 0 & \cos\beta \end{bmatrix} \begin{bmatrix} \cos\gamma & -\sin\gamma & 0 \\ \sin\gamma & \cos\gamma & 0 \\ 0 & 0 & 1 \end{bmatrix} \begin{bmatrix} \Delta x_{m1} \\ \Delta y_{m1} \\ \Delta z_{m1} \end{bmatrix}$$

$$(5\text{-}22)$$

Assuming that the feature m_1 has an initial manufacturing error of ξ_{m_1}, taking the error $\Delta \boldsymbol{m}'$ into account, the error of m_1 at the kth assembly station should be modified according to the practical assembly situation, and the modification value can be calculated with Equation (5-23).

$$\xi'_{m_1} = \xi_{m_1} + \Delta m' \qquad (5\text{-}23)$$

In Figure 5-8, where CS, $O_2X_2Y_2Z_2$, represents the manufacturing datum of part 2. Combined with the references of part 1, they are assembled in another assembly CS at the kth assembly station, namely, $O_kX_kY_kZ_k$. To get the position error of the key features on part 2, the transformation relationship from the CSs of $O_1X_1Y_1Z_1$ and $O_2X_2Y_2Z_2$ to the unified assembly CS of $O_kX_kY_kZ_k$ should be analyzed firstly, and a homogeneous matrix is derived. Then according to the manufacturing accuracy of features on part 2 and the position relationships under CS, $O_2X_2Y_2Z_2$, spatial position error of the key features at current assembly station can be calculated with Equation (5-4) that mentioned in Section 5.3.1.

At the $(k+1)$th assembly station, the feature's relocation error is caused by the changes of datum features that distributed on different parts. To analyze the error changes at different assembly stations, the CSs should be unified to the CS, $O_{k+1}X_{k+1}Y_{k+1}Z_{k+1}$, for the entire component firstly. By referencing ① the basic manufacturing errors of part 2 and part 3, and ② the spatial position relationships in $O_{k+1}X_{k+1}Y_{k+1}Z_{k+1}$, taking the error relationships among the key features on part 1 and part 2 into account, the relocation error C_{ij} of the key features at current assembly station on part 1 can be gained based on Equation (5-4) as well.

5.5 Coordination Error Chain Construction Based on Error Propagation Modeling

Aircraft manufacturing is a progressive assembly process of parts, components, and the whole aircraft, which are assembled using different kinds of tooling or jigs. In order to facilitate the production, assembly stations are often mentioned and divided as the decomposition and organization of the complex manufacturing process. In practical site, a station may contain each or more set of jigs/fixtures. However, to construct coordination error chain conveniently, it is assumed that each station is in correspondence with only one assembly jig in this section. To the assemblies that have a coordination error relationship, there are two kinds of propagation processes related to assembly errors: the single assembly station and the multiple assembly stations, as modeled in the next two sections.

5.5.1 Error Propagation for Single Assembly Station

The assembly work at single assembly station is an accumulation process for parts to be assembled on the semi-finished assemblies, and the assembly process follows a certain

Digital Assembly Coordination and Quality Controlling Technology
for Aeronautical Thin-walled Structures

given sequence that has a relationship with error propagation process. The accumulation and transformation process of the comprised parts' spatial positions would be demonstrated with an error propagation chain, which is comprised of several kinds of error items. The propagating process can be taken as the construction of an error chain for coordination error between different assemblies. For a given assembly station k, the error propagation chain can be shown in Figure 5-9.

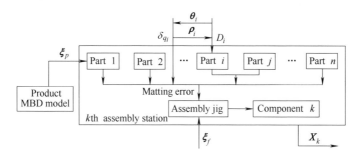

Figure 5-9   Error propagation process for one single assembly station

In the error propagation process, assembly error of the key features has three main sources: manufacturing error $\xi_p(k)$ of parts, installation and positioning error $\xi_f(k)$ of assembly tooling, and deformation error $\xi_d(k)$ during assembly process. However, interaction of the above three kinds of error would bring the following assembly errors: mating error $\delta_q(k)$ caused by $\xi_p(k)$ and $\xi_f(k)$, influence error $\theta_i(k)$ caused by $\delta_q(k)$ on $\xi_d(k)$, and influence error $\rho_i(k)$ caused by $\xi_d(k)$ on $\delta_q(k)$. Based on the above six kind of errors, without considering other random errors and the influence effects of the errors at previous assembly station, the accumulated assembly error $[X_k]_0$ of the key features on final component can be expressed by Equation (5-24). It is mentioned that this establishment process can be taken as one segment of coordination error chain.

$$[X_k]_0 = F(\xi_p(k), \xi_f(k), \xi_d(k)) = \delta_q(k) + \xi_d(k) + \theta(k) + \rho(k) \qquad (5\text{-}24)$$

5.5.2   Error Propagation for Multiple Assembly Stations

As different assembly processes and stations flowing down, the errors would propagate and accumulate on different assemblies to form the final assembly error of the whole product. However, the analysis on error propagation for multiple assembly stations is not the same as the single assembly station, due to the existences of other error items, as shown in Figure 5-10.

Assuming that there are n assembly stations for one component's assembly work. For the first assembly station, m parts are assembled on the jig. And for the assembly

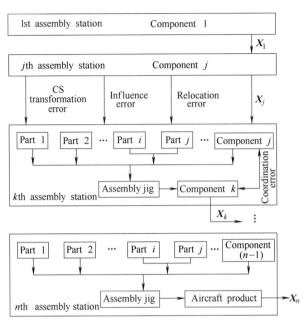

Figure 5-10   Error propagation process for multiple assembly stations

work of two adjacent assembly stations j and k, there often exists a phenomenon that a component j at the previous assembly station acts as a special part for the next station. Considering the measurement data and analysis work for an assembly process are often based on the CS of tooling, when assembly error of component k is analyzed, CS of assembly jigs, C_j, at the jth assembly station should be transformed to CS, C_k, at the kth assembly station firstly, to determine the spatial position of component j. This CS transformation would bring an error item, $T_{i,j}$. Then, considering the change of locating datum at current assembly station, the relocation process, where component j is to be assembled at the kth assembly station with different datum, would bring a new relocation error, which is donated as $C_{i,j}$. In addition, deformation error of component j would act as an input error item for assembly error at the kth assembly station. Meanwhile, spatial position error of component j that caused by the mating error in C_k would also influence the following deformation error at the same assembly station.

Assuming that the assembly work of the jth and the kth assembly stations has a close precedence relationship, then assembly error X_{k-1}, i.e., X_j, can be taken as a comprised input error item or an error link for the kth assembly station. Taking the influence error that caused by assembly error at previous assembly station into account, assembly error at the current kth assembly station can be modeled by Equation (5-25). Where, $[X_k]_0$ stands for assembly error of the kth assembly station without considering coordination

Digital Assembly Coordination and Quality Controlling Technology
for Aeronautical Thin-walled Structures

datum transformation, and $\varLambda_{k-1,k}$ represents the transformation and influence errors between jth and the kth assembly stations.

$$
\begin{aligned}
X_k &= [X_{k-1}]_0 + [X_k]_0 + \varLambda_{k-1,k} \\
&= [X_{k-1}]_0 + F(\boldsymbol{\xi}_p(k), \boldsymbol{\xi}_f(k), \boldsymbol{\xi}_d(k)) + (T_{k-1,k} + C_{k-1,k} + \Delta CP_{k-1,k} + \Delta CT_{k-1,k}) \\
&= [\boldsymbol{\delta}_q(k-1) + \boldsymbol{\xi}_d(k-1) + \boldsymbol{\theta}(k-1) + \boldsymbol{\rho}(k-1)] + [\boldsymbol{\delta}_q(k) + \boldsymbol{\xi}_d(k) + \boldsymbol{\theta}(k) + \boldsymbol{\rho}(k)] \\
&\quad + (T_{k-1,k} + C_{k-1,k} + \Delta CP_{k-1,k} + \Delta CT_{k-1,k})
\end{aligned}
$$

$$(5\text{-}25)$$

To facilitate application for coordination error chain modeling, several interruptions are stated here based on Equation (5-25) for different assemblies in multi-hierarchy assembly process.

① In the dimension and shape transfer processes, if two random geometric features CE_p and CE'_p that distributed on different parts have a coordination relationship with each other, and are manufactured according to their corresponding theoretical MBD models, then, according to the transformation relationship from part's manufacturing CSs to an unified assembly CS, then the coordination error chain can be calculated with Equation (5-26). Where, ∇_{CE_p,CE'_p} represents the coordination error between the two corresponding assembly features, CE_p and CE'_p, T_A and T'_A stand for the transformation matrices, A_p and A'_p represent the original spatial positions in their manufacturing CSs, while $\boldsymbol{\xi}_{CE_p}$ and $\boldsymbol{\xi}_{CE'_p}$ represent manufacturing errors of the above features.

$$
\nabla_{CE_p,CE'_p} = T_A \cdot A_P - T'_A \cdot A'_P + \boldsymbol{\xi}_{CE_p} - \boldsymbol{\xi}_{CE'_p} \tag{5-26}
$$

② To the coordination error relationship between a part and a component, assuming that the manufacturing error for part's feature is donated as $\boldsymbol{\xi}_{CE_p}$, and two coordination features are CE_p and CE_c that located on final component. Given there are k continuous assembly stations to form the dimension and shape errors of the feature CE_c, then the final assembly error $\Delta X_{CE_c}(k)$ at the kth assembly station can be modeled by Equation (5-27). Where, $\varLambda_{i,j}$ represents the transformation error as the coordination datum changes from the ith assembly station to jth, with $i<j<k$ and $k \geqslant 1$.

$$
\begin{aligned}
\Delta X_{CE_c}(k) &= [X_1]_0 + [X_2]_0 + \cdots + [X_{k-1}]_0 + [X_k]_0 \\
&\quad + \varLambda_{1,2} + \varLambda_{2,3} + \cdots + \varLambda_{k-1,k} = \sum_{i=1}^{k}[X_i]_0 + \sum_{i=2}^{k}\varLambda_{k-1,k}
\end{aligned}
\tag{5-27}
$$

Then, the coordination error chain for the above situation can be calculated with Equation (5-28).

$$\nabla_{CE_p CE_c} = \xi_{CE_p} + T_B \cdot A_P - \Delta X_{CE_c}(k)$$

$$= \xi_{CE_p} + T_B \cdot A_P - \sum_{i=1}^{k}[X_i]_0 - \sum_{i=2}^{k}\Delta_{k-1,k} \qquad (5\text{-}28)$$

③ To the situation that two coordination features distributing on different components, assuming that the components are assembled with k continuous assembly stations, as shown in Figure 5-11.

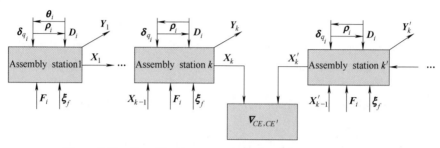

**Figure 5-11   Coordination error chain for different components
with k continuous assembly stations**

By calculating the accumulated assembly error $\Delta X_{CE}(k)$ and $\Delta X_{CE'}(k)$ at the above two stations, then the coordination error chain for different components during assembly process can be constructed, and the coordination error $\nabla_{CE,CE'}$ is given by Equation (5-29).

$$\nabla_{CE,CE'} = \Delta X_{CE}(k) - \Delta X_{CE'}(k) = \sum_{i=1}^{k}[X_i]_0 + \sum_{i=2}^{k}\Delta_{i-1,i} - \sum_{i=1}^{k}[X_i']_0 - \sum_{i=2}^{k}\Delta_{i-1,i}' \quad (5\text{-}29)$$

④ To the situation that assembly error X_{k-1} at the previous assembly station is not an input error item for the current assembly station, i.e., the error X_{k-1} has no relationship with the current accumulation error X_k, and the situation that there is no coordination relationship, i.e., there is no coordination errors for shape and dimension, between different assemblies at adjacent assembly stations, then assembly errors X_{k-1} and X_k are only taken as error links in coordination error chain, and they are just related to the error propagation process for a single assembly station.

5.5.3   Discussion

To ensure the consistency of the final dimension/shape/position errors, when the accumulated assembly errors on different assemblies are not in harmony with each other, a feedback adjustment work should be done to the error links of those assemblies. Then

Digital Assembly Coordination and Quality Controlling Technology
for Aeronautical Thin-walled Structures

the controlling method for coordination accuracy can be interpreted by Figure 5-2. Considering that if all comprised error items for assembly error $\Delta X_{CE'}(k)$ at the all k assembly stations are adjusted, then the improved accumulated error $\Delta X''_{CE}(k)$ under this extreme situation at the kth assembly station is modeled by Equation (5-30).

$$\Delta X''_{CE}(k) = \sum_{i=1}^{k} [X''_i]_0 + \sum_{i=2}^{k} \Delta''_{i-1,i} \qquad (5\text{-}30)$$

For one given assembly station i, if the adjusted error items at the ith assembly station include ① mating error $\delta_q(i)''$, ② assembly deformation error $\xi_d(i)''$, and ③ influence errors of $\theta(i)''$ and $\rho(i)''$ that caused by the error changes of $\delta_q(i)'$ and $\xi_d(i)'$, then the assembly error can be expressed with Equation (5-31).

$$[X''_i]_0 = [\delta_q(i)'' + \xi_d(i)'' + \theta(i)'' + \rho(i)''] \qquad (5\text{-}31)$$

Then to the error $\Delta''_{i-1,i}$ that caused in the transformation process for coordination datum, its deviation can be calculated by Equation (5-32). Where the adjusted errors are CSs transformation error $T''_{i-1,i}$, and the errors items of $\Delta CP''_{i-1,i}$ and $\Delta CT''_{i-1,i}$.

$$\Delta''_{i-1,i} = T''_{i-1,i} + C''_{i-1,i} + \Delta CP''_{i-1,i} + \Delta CT''_{i-1,i} \qquad (5\text{-}32)$$

With the feedback adjustment on the error items, the coordination error chain between two different assemblies that shown in Figure 5-11 can be modeled by Equation (5-33). Lastly, based on the calculation results, a comparing work can be done with the required assembly coordination accuracy to judge whether the accumulated coordination errors are in harmony with each other or not.

$$\nabla_{CE',CE''} = \Delta X_{CE'}(k) - \Delta X_{CE''}(k) = \sum_{i=1}^{k} [X'_i]_0 + \sum_{i=2}^{k} \Delta'_{i-1,i} - \sum_{i=1}^{k} [X''_i]_0 - \sum_{i=2}^{k} \Delta''_{i-1,i} \quad (5\text{-}33)$$

5.6 Experimental Verification

In this section, the assembly work of a wing component would be taken as the experiment object to verify the feasibility of the proposed method. The whole wing component consists of four individual leading edge flap components and the main body of the wing component. At the station for assembling leading edge flap components, to ensure the profile accuracy of the skin and rib, assembly datum of the center rib is the locating hole and the manipulating joint of the rib. The above two datums are connected by a black line, and the detailed comprised parts are as also shown in Figure 5-12. When the assembly work of single component is completed, the component is then transferred to another assembly station, where it is joined with the wing component directly without special locaters in the finishing jig. However, at this assembly station, the locating datum of the leading edge flap component is the manipulating joint of the rib and the hinge joint of the spar, which are connected by a gray line in the figure.

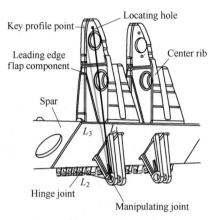

Figure 5-12   Transformation of the coordination datum in the jointing process
(See the color illustration)

By measuring the theoretical positions of these features in design coordinate system of the whole aircraft, coordinates of the manipulating joint's center is (7133.859, 324.525, −2002.634)mm, the hinge joint's center is (7100.088, 224.208, −1992.087)mm, and coordinates of the special key profile point that shown in the figure is (6720.704, 254.973, −2189.583)mm. The distance L_1 between the manipulating joint and the locating hole is 440mm; the distance L_2 between the manipulating joint and the hinge joint is 110mm; the distance L_3 between the manipulating joint and the key profile point is 460mm. In this assembly transformation case, the main comprised error items are analyzed as follows.

① In the assembly process of leading edge flap component, a measuring work is done to record the practical accuracy status with laser tracker (API track III). Along the specific normal direction of the profile surface, it can be known that the manipulating joint's positioning error is 0.05mm, and the locating hole's positioning error is 0.04mm. It is mentioned that the precision value along the normal direction of skin profile is the judging basis for determining whether the assembly accuracy is satisfied or not. Then according to the spatial position relationship and the practical assembly physical state of the current component, with the interaction relationship of basic error items in Section 5.3, the calculated assembly precision Δ_{l-e-f} of the key profile point that distribution on the leading edge flap component is 0.31mm. However, when the assembly operation work is done, the practical measured precision result of Δ_{l-e-f} is 0.28mm, and it is almost in accordance with the theoretical calculation result.

② As the geometric features of the leading edge component and the wing component are designed in the unified aircraft CS, correspondingly, there is no transformation error for each component's design CS. But for the finishing jig used to assemble the whole wing component, its CS is built on the basis of three TB (tooling ball) points: TB_1^0, TB_2^0, and TB_3^0. Their theory spatial positions are (5833.766, −1392.992, −1890.015)mm, (9933.774, −1392.992, −1890.015)mm, and (9933.847, −1393.014, −5550.085)mm, respectively. These three TB points cover the main structure of the finishing jig to the utmost.

However, as building the CS in practical, the actual spatial positions of the three TB points, i.e., TB_1', TB_2', and TB_3', are (5833.780, −1393.000, −1890.000)mm, (9933.750, −1393.000, −1890.000)mm, and (9933.820, −1393.000, −5550.070)mm, respectively. Then, the detailed deviation of the three datum points, i.e., ΔTB_1, ΔTB_2, and ΔTB_3, can be calculated as (−0.014, 0.008, −0.015)mm, (0.024, 0.008, −0.015)mm, and (0.027, −0.014, −0.015)mm, respectively. Based on the above data, for the finishing assembly jig, the transformation relationship between the theory assembly CS and the practical assembly CS can be donated by Equation (5-34) and (5-35).

$$M_T = \begin{bmatrix} 1 & 0 & 0 & 0.024 \\ 0 & 1 & 0 & 0.008 \\ 0 & 0 & 1 & -0.015 \\ 0 & 0 & 0 & 1 \end{bmatrix} \tag{5-34}$$

$$M_R = \begin{bmatrix} 1 & 0 & 0 & 0 \\ 0 & \cos 0 & -\sin 0 & 0 \\ 0 & \sin 0 & \cos 0 & 0 \\ 0 & 0 & 0 & 1 \end{bmatrix} \begin{bmatrix} \cos 0.015 & 0 & \sin 0.015 & 0 \\ 0 & 1 & 0 & 0 \\ -\sin 0.015 & 0 & \cos 0.015 & 0 \\ 0 & 0 & 0 & 1 \end{bmatrix} \begin{bmatrix} \cos 0.015 & -\sin 0.015 & 0 & 0 \\ \sin 0.015 & \cos 0.015 & 0 & 0 \\ 0 & 0 & 1 & 0 \\ 0 & 0 & 0 & 1 \end{bmatrix}$$

$$\tag{5-35}$$

By measuring the theory position of the key profile point (as shown in Figure 5-13) in the finishing assembly jig, it has a spatial position of (−3213.070, 1647.965, −299.568)mm, which is donated as y. With the Equation (5-34) and (5-35), the actual position deviation that caused by the transformation error (unit: mm) between the practical CS and the theory CS is given by Equation (5-36).

$$\Delta y = y' - y = M_T + M_R \cdot y - y = [-0.374, -0.833, -0.822]^T \tag{5-36}$$

With the analysis of the above result, it can be known that the position error of three TB points has a great impaction on the accuracy of the product profile. The maximum error Δy of the key profile point is 0.833mm at y-direction, and this deviation is unacceptable according to the official assembly document, because it even exceeds the required assembly accuracy of [−0.6, +0.4]mm.

To decrease this enormous influence, a solution of adding relevant ERS points around the key profile point on finishing jig is put into application. By improving the positioning accuracy of TB points and ERS points, a new CS of the finishing jig is established for the assembling site based on the practical measuring data. With the Equation (5-34) and (5-35), the spatial position error of the key profile point that caused by CS transformation is calculated with Equation (5-37) (unit: mm), and this result is acceptable in actual engineering. In summary, the above solution, building an accurate assembly tooling CS according to the impact on key assembly features, is the first step to guarantee coordination error at the current assembly station.

$$\Delta y' = y'' - y_{ERS} = M_T + M_R \cdot y_{ERS} - y_{ERS} = [0.070, 0.092, 0.090]^T \qquad (5-37)$$

③ At the current assembly station, to the influence error θ_i of mating error between the two components on ensuing assembly deformation, this influence error item can be ignored, as the two components are only jointed together with each other without the detailed subsequent clamping-jointing-releasing stage.

④ At the present assembly station, the leading edge flap component is jointed directly with the wing component, and the hinge joints of wing component can be taken as specific end locators for the leading edge flap component. As a result, the coordination error between the "tooling" of these two assembly stations can be taken as the mating error of different hinge joints, and the error along the normal direction of the profile surface can be calculated according to Equation (5-3) and (5-4). When the hinge joints of the wing are located on the finishing assembly jig, a measuring work is done, and their detailed position accuracy Δ_1' and Δ_2' are 0.05mm. And the profile accuracy Δ_w of the wing component has a deviation range of [−0.60, 0.40]mm. Then, along the normal direction of skin profile, to the area of key corresponding profile points on the wing component, the measured result is 0.33mm. However, for the leading edge flap component at the current assembly station, the practical measured assembly precision Δ_1 of the hinge joint is 0.06mm, and assembly precision Δ_2 of the manipulating joint is 0.10mm. The above results can be taken as the comprehensive error items of the influence error of assembly deformation at the previous assembly station on current station, and the mating error at the current station. However, without considering the influence effort, the assembly deformation error items of Δ_1^0 and Δ_2^0 are 0.042mm and 0.087mm, respectively.

When the hinge joints and the manipulating joints of the leading edge flap component and the wing component mate with each other, they have a mating relationship as shown in Figure 5-13. With the consideration of position error Δr^{pos} and dimension error Δr^{size} of these locating holes, the comprehensive error deviation $\Delta r_{hole-pin}^{h}$ for the hinge joint is calculated by the Equation (5-3) and (5-4), having a range of [−0.01, 0.12]mm, and the deviation $\Delta r_{hole-pin}^{m}$ for the manipulating joint has a range of [−0.04, 0.14]mm.

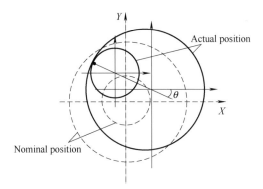

Figure 5-13   Matting relationship for the joint holes of different components

Based on the analysis of Section 5.4.3, a relocation error would occur due to the transformation of locating datum for the leading edge component. According to the positional and dimensional relationship of the locating hole and the joints, the rotation angle $\Delta\alpha$ of the leading edge flap component around the axis of manipulating joint at the present assembly station is given by Equation (5-38).

$$\Delta\alpha \approx \Delta / L_1 = (\Delta r_{hole-pin}^{h} + \Delta r_{hole-pin}^{m}) / L_1 = [-0.05, 0.26] / 110 \qquad (5\text{-}38)$$

With the distance relationship of the key profile point and the manipulating joint, the deviating range ΔS of the key profile point, which is perpendicular to the wing datum plane and caused by $\Delta\alpha$, can be calculated with Equation (5-39) (unit: mm).

$$\Delta S = \Delta\alpha \times L_3 = \frac{\Delta}{L_1} \times L_3 = \frac{[-0.05, 0.26]}{110} \times 460 = [-0.21, 1.08] \qquad (5\text{-}39)$$

Based on the above analysis, the range of coordination error for the key profile point on the leading edge flap component and the skin profile of the wing component, which is donated as ∇_{w-lef}, can be calculated by Equation (5-40) (unit: mm).

$$\nabla_{w-lef} = \Delta_w - (\Delta y' + \Delta_{l-e-f} + \Delta S) = [-0.258, 1.112] \qquad (5\text{-}40)$$

Without considering the influence effort, i.e., only the assembly deformation error items of the previous station are taken into account, coordination error of the key profile point can be calculated by Equation (5-41) (unit: mm).

$$V'_{w-lef} = \Delta_w - (\Delta y' + \Delta_{t-e-f} + \Delta S^0) = [-0.177, 0.784] \quad (5\text{-}41)$$

Then according to results in Equation (5-40), it can be easily known that the coordination error is out of the required range of 1mm, although it is in accordance with the practical assembly site. However, in Equation (5-41), the result is within the accuracy requirement, and there is a bad conformance with reality, due to the neglect of the influence effort of assembly deformation on mating error at different assembly stations. The inconsistency phenomenon is the same as observed without considering the rotation error that calculated in Equation (5-38) and (5-39).

In practical assembly site, by analyzing the compositions of the error, it is easily to find that the inharmonic phenomenon is mainly caused by the error item ΔS that listed in Equation (5-39). A solution work is carried out to optimize the comprised error items for the hinge joint and manipulating joint. With the optimized error items, the calculated comprehensive error deviation $\Delta r^h_{hole-pin}{}'$ for the hinge joint has a range of $[-0.05,$ $0.06]$mm, and the comprehensive error deviation $\Delta r^m_{hole-pin}{}'$ for the hinge joint has a range of $[-0.04, +0.07]$mm. Then, the optimized result $\Delta\alpha' \approx \Delta'/L_1 = [-0.09, 0.13]/110$, and the corresponding deviation $\Delta S'$ has a range of $\Delta S' = \Delta\alpha' \times L_3 = \dfrac{\Delta'}{L_1} \times L_3 =$ $\dfrac{[-0.09, 0.13]}{110} \times 460 = [-0.37, 0.54]$mm. Finally, the coordination error has an optimized range of $V''_{w-lef} = \Delta_w - (\Delta y' + \Delta_{t-e-f} + \Delta S') = [-0.418, 0.572]$mm, and this theoretical calculation result can satisfy the flush accuracy requirement of 1mm.

To verify the methodology's feasibility, by controlling the direction and the value of error range for the hinge joint and manipulating joint, a measuring work is done. The measured results can be shown in Table 5-1. Where the positive values indicate that the profile detection results exceed the theoretical requirement, and the negative values indicate a smaller profile. From this, we can know the maximum value of the practical flush is 0.681mm, the minimum value is only 0.021mm, and the average flush of the entire wing component is 0.358mm, which are in accordance with theoretical calculation results of 0.572mm and can successfully meet the assembly requirement.

Table 5-1   The measured practical profile accuracy

Error items	Precision/mm	Error items	Precision/mm
Leading edge flap component	0.394	Wing component	0.287
	0.344		0.153
	0.238		−0.373
	−0.327		0.214

5.7 Summary

① According to the error propagation process when assembling aircraft parts or components, based on the three basic kinds of error sources, i.e., the manufacturing error of parts, the positioning error of assembly tooling, and the assembly deformation error, the interaction effects of the mating error, the deformation error, along with their mutual influence relationship, are modeled as the assembly process flowing down.

② Since the assembly error have a relationship with the dimension error and the position error, with the characteristics of multi-hierarchy and multi-station for the assembly work of an aircraft, the coordination datum transformation is defined. The establishment and transformation errors of coordinate systems, the relocation error caused by the changes of locating datum, and the coordination error between two mating objects are modeled in the assembly process. Then the actual positions of aircraft parts or components in the unified coordination system of aircraft are modified and determined to make the calculated assembly error more accurate.

③ Taking the progressive assembly process of parts, components, and the whole aircraft into account, the error propagation for a single assembly station and for multiple assembly stations is modeled to gain coordination error chain for different assemblies, and the final coordination error is modeled.

④ For the assembly error propagation process across different assembly stations, the verified objects are the leading edge flap component and the wing component at different assembly stations. By constructing the coordination dimension chain, and with certain optimized solutions, the maximum value of the practical flush of the profiles between the two components is only 0.681mm, which is in accordance with theoretical calculation results and can successfully meet the assembly requirement.

References

[1] Bullen G. Automated/mechanized drilling and countersinking of airframes[M]. Warrendale: SAE International Press, 2013.

[2] Jin Y, Abella R, Ares E, et al. Modeling and digital tool development of a new similarity metric for aerospace production[J]. International Journal of Advanced Manufacturing Technology, 2013, 69(1-4): 777-795.

[3] Williams G, Chalupa E, Billieu R, et al. Gaugeless tooling[R]. SAE Technical Paper, 1998.

[4] Quintana V, Rivest L, Pellerin R, et al. Will Model-based Definition replace engineering drawings throughout the product lifecycle? A global perspective from aerospace industry[J]. Computers in Industry, 2010, 61(5): 497-508.

[5] Muelaner J, Cai B, Maropoulos P. Large-volume metrology instrument selection and measurability analysis[J]. Proceedings of the Institution of Mechanical Engineers Part B Journal of Engineering Manufacture, 2010, 224(6): 853-868.

[6] Grieves M, Vickers J. Digital twin: Mitigating unpredictable, undesirable emergent behavior in complex systems[M]. Berlin: Springer International Publishing, 2017.

[7] Aehnelt M, Bader S. Tracking assembly processes and providing assistance in smart factories[J]. ICAART, 2014: 161-168.

[8] Guo F, Zou F, Liu J. Working mode in aircraft manufacturing based on digital coordination model[J]. The International Journal of Advanced Manufacturing Technology, 2018, 98: 1547-1571.

[9] Phoomboplab T, Ceglarek D. Design synthesis framework for dimensional management in multistage assembly system[J]. CIRP Annals-Manufacturing Technology, 2007, 56(1): 153-158.

[10] Huang W, Kong Z. Simulation and integration of geometric and rigid body kinematics errors for assembly variation analysis[J]. Journal of Manufacturing Systems, 2008, 27(1): 36-44.

[11] Saravanan A, Balamurugan C, Sivakumar K, et al. Optimal geometric tolerance design framework for rigid parts with assembly function requirements using evolutionary algorithms[J]. International Journal of Advanced Manufacturing Technology, 2014, 73(9-12): 1219-1236.

[12] Li F, He H, Li P, et al. Tolerance analysis of mechanical assemblies based on small displacement torsor and deviation propagation theories[J]. The International Journal of Advanced Manufacturing Technology, 2014, 72(1-4): 89-99.

[13] Guo C, Liu J, Jiang K. Efficient statistical analysis of geometric tolerances using unified error distribution and an analytical variation model[J]. International Journal of Advanced Manufacturing Technology, 2016, 84(1-4): 347-360.

[14] Franciosa P, Gerbino S, Patalano S. Variational modeling and assembly constraints in tolerance analysis of rigid part assemblies: Planar and cylindrical features[J]. International Journal of Advanced Manufacturing Technology, 2010, 49(1-4): 239-251.

[15] Liu J, Jin J, Shi J. State space modeling for 3-d variation propagation in rigid-body multistage assembly processes[J]. IEEE Transactions on Automation Science & Engineering, 2010, 7(2): 274-290.

[16] Du S, Yao X, Huang D, et al. Three-dimensional variation propagation modeling for multistage turning process of rotary workpieces[J]. Computers & Industrial Engineering, 2015, 82(C): 41-53.

[17] Bakker O, Popov A, Ratchev S. Variation analysis of automated wing box assembly[J]. Cirp Conference on Manufacturing Systems, 2017, 63: 406-411.

[18] Das A, Franciosa P, Prakash P, et al. Transfer function of assembly process with compliant non-ideal parts[J]. Procedia Cirp, 2014, 21: 177-182.

[19] Franciosa P, Gerbino S, Lanzotti A, et al. Automatic evaluation of variational parameters for tolerance analysis of rigid parts based on graphs[J]. International Journal on Interactive Design&Manufacturing, 2013, 7(4): 239-248.

Digital Assembly Coordination and Quality Controlling Technology
for Aeronautical Thin-walled Structures

[20] Cai N, Qiao L. Rigid-compliant hybrid variation modeling of sheet metal assembly with 3D generic free surface[J]. Journal of Manufacturing Systems, 2016, 41(3): 45-64.

[21] Camelio J, Hu S, Marin S. Compliant assembly variation analysis using component geometric covariance[J]. Journal of Manufacturing Science & Engineering, 2014, 126(2): 355-360.

[22] Lin J, Jin S, Zheng C, et al. Compliant assembly variation analysis of aeronautical panels using unified substructures with consideration of identical parts[J]. Computer-Aided Design, 2014, 57:29-40.

[23] Qu X, Li X, Ma Q, et al. Variation propagation modeling for locating datum system design in multi-station assembly processes[J]. The International Journal of Advanced Manufacturing Technology, 2016, 86(5-8): 1357-1366.

[24] Wang Q, Hou R, Li J, et al. Positioning variation modeling for aircraft panels assembly based on elastic deformation theory[J]. SAGE Publications, 2018, 232(14): 2592-2604.

[25] Li Y, Zhao Y, Yu H, et al. Compliant assembly variation analysis of sheet metal with shape errors based on primitive deformation patterns[J]. Proceedings of the Institution of Mechanical Engineers, Part C. Journal of Mechanical Engineering Science, 2018, 232(13): 2334-2351.

[26] Wang Q, Dou Y, Li J, et al. An assembly gap control method based on posture alignment of wing panels in aircraft assembly[J]. Assembly Automation, 2017, 37(1): 422-433.

[27] Xie K, Camelio J A, Izquierdo L. Part-by-part dimensional error compensation in compliant sheet metal assembly processes[J]. Journal of Manufacturing Systems, 2012, 31(2): 152-161.

[28] Cheng H, Li Y, Zhang K, et al. Variation modeling of aeronautical thin-walled structures with multi-state riveting[J]. Journal of Manufacturing Systems, 2011, 30(2): 101-115.

[29] Wang H. Investigation on the effect of spring-in distortion on strength of a biomaterial beam[J]. Journal of Aerospace Engineering, 2016, 29(3): 04015069.

Chapter 6
Assembly Quality Control and Reliability improvement with Feedback Actions

6.1 Introduction and Related Work

For complex and major products, such as aircraft and automobile, the main structures and other core bearing components are usually riveted/welded/screwed/glued together using a large number of thin-walled parts that have a low rigidity[1]. Since the assembled structures would endure complex alternating and impacting loads during their service period, for the numerical size and distribution status of the dimensional accuracy and internal stress after assembly, they will affect the service indicators directly, such as product's quality, fatigue, and damage performance[2,3]. For example, for aircraft products, ① assembly accuracy of skin profile affects the aerodynamic parameters, ② installation accuracy of movable wing components and manipulation mechanisms affects the maneuverability indices, ③ step difference and butting gap between fixed skin and other movable parts affect the stealth performance, ④ positioning and joining quality, and the residual stress and damage generated during assembly process, affect the mechanical performance, bearing strength and fatigue life of the assembled components, as well as other structure integrity indicators (according to statistics, 80% of the fatigue damage of aircraft's structure occurs at the joining area[4]), ⑤ geometric accuracy of outline profile and position of reference axis also affect the production interchangeability of the comprised large components.

As the final link of precision controlling in the development of light-weighted and thin-walled structures, assembly is relevant with other process links, such as product design, part manufacturing, assembly process planning, tooling-assisted positioning, hole's drilling and joining, measurement and quality evaluation, and it accounts for about 30%~60% of the workload as manufacturing complex products. Controlling assembly quality presents the following difficulties, such as: ① huge number of parts with large

Digital Assembly Coordination and Quality Controlling Technology
for Aeronautical Thin-walled Structures

size and low rigidity, and tight restriction for aerodynamic shape (smooth streamline is often required), ② lengthy assembly procedures with a quantity of error links that make the error transmission path extremely complex, ③ many types of dedicated assembly fixture/tooling/jigs are adopted[5], whose accuracy is usually 3~5 times higher than the accuracy of product parts, ④ the general Tolerance and Fit system is inadequate to ensure the final assembly accuracy, and the support of complex and robust interchange/coordination theory is essential. For the current assembly theory based on geometric quantity controlling, it has the typical characteristics of "digital/empirical, passive repair, and matching on geometric shape". It is an open-loop assembly system without feedback improvements. Correspondingly, assembly accuracy is guaranteed mainly by compensation operations and worker's personal ability[6,7], indicating the structure's performance cannot be guaranteed effectively and efficiently. In conclusion, the current assembly methods lack the ability to ensure that the actual assembly accuracy meets design requirements, which means the reliability of assembly accuracy should be enhanced.

With the development of digital, automatic and intelligent technologies, design level and manufacturing consistency of major equipment have been improved significantly[8]. Assembly locates at trailing end of production, and the performance guarantee of assembled structures is gradually transferred from initial design and machining links to assembly links. As a result, assembly technology is highly concerned[9-11]. As a matter of fact, when assembling thin-walled structures, final accuracy is the superposition of initial geometric error and internal physical stress that generated in the entire assembly process. To guarantee structures' performance, firstly, it is necessary to reduce the occurring probability of out-of-tolerance phenomenon with optimization and adjustment measures[12] in advance. To be more specific, considering the nonlinear accumulation of different error sources, this phenomenon can be called as EOCDCF (extreme out-of-tolerance caused by deviation's coupling and focusing function) event. In fact, it's difficult to improve accuracy reliability for thin-walled structures, for the following two aspects: ① nonlinear coupling and focusing effect among multi-dimensional error sources make it difficult for the assembly geometric accuracy to meet design requirements, which means the EOCDCF event has a high occurring probability, ② on practical engineering site, it is also often impossible to realize accurate repair compensation in advance, only manual trial and error methods are adopted to ensure geometric accuracy, resulting in low efficiency and excessive structure's internal stress.

Enhancing the assembly accuracy, is also one of the enduring goals of product's manufacturing. Considering the disturbance of multi-dimensional error sources, assembly error modeling based on rigid hypothesis is relatively mature, and can represent the accurate motion deviation. However, when describing deviation disturbance caused by deformation of complex thin-walled structures, such as skin panel components, it is not suitable anymore. To enhance predictive accuracy of assembly quality, with actual deformation and geometric manufacturing errors, Kang[13] modified and extended SDT (small displacement tensor) expression of Jacobian-Tensor model. Then by updating mating constraints of actual flexible parts, Tlija[14] conducted finite element simulations with the distribution state of measured errors, and Ballu[15] analyzed the nonlinear accumulation relationships with physical characteristics and optimization constraints (assembly loads, displacement boundary conditions). In addition, considering: ① mating status of non-ideal surfaces, and ② external and internal loads, methods such as linear complementary were also proposed, leading to the conclusion that assembly tolerance analysis with SMS (skin model shape) models could be taken as the objective function of a quadratic optimization problem. With DT (digital twin) solutions, Yi[16] calculated assembly errors with updating and iterating mechanism. In addition, by integrating 3D geometric model and massive measurement data, Sun[17] and Masnaoui[18] proposed "semi-physical assembly" technology, which could improve the success rate of one-time assembly. However, for the geometric-physical field coupling and errors' multi-scale transmission process, it may leads to the amplification and focusing effort, i.e., assembly deviation exceeds design expectation, then, unstable and serious quality problems could occur. From another perspective, although the finite element simulation are adopted, it is still difficult to effectively use on-site measured data to reduce assembly uncertainty, preventing the accurate guidance on executing adjustment instructions.

Probability modeling of assembly EOCDCF event can quantify the accumulation results of different error items and reduce the influence of uncertainty factors, such as worker's misoperations. Therefore, analysis of probability under the disturbance of multi-dimensional error sources, can prevent its occurrence. LDP (large deviation principle) is an important portion of probability theory to describe the occurrence of rare events[19,20]. Then, considering the assembly process of complex thin-walled structures, by comparing the probability value of current assembly state with defined probability threshold, the reliability of assembly accuracy can be quickly evaluated. Considering error sources in assembly process should be regarded as probability variables with an independent and identically distribution, LDP is applicable to estimate the probability of EOCDCF event, which means it can realize the application of probability method to improve accuracy reliability, as well as predict serious quality problems. For parts with free-form surfaces, Corrado[21] investigated the influence of tail beams' errors on the gaps among structure interfaces. Where, the fitting condition and tolerance were modeled with probability density function. With rigid-flexible model, the tolerance zone and

Digital Assembly Coordination and Quality Controlling Technology
for Aeronautical Thin-walled Structures

distribution of target assembly deviations could be obtained. Stricher[22] improved the influence coefficient method by considering dynamic physical characteristics. Where, the geometric nonlinearity caused by stiffness changes was considered, and the coupling results of joining defects, shape tolerances, and rigidity changes on final assembly deviation for beam component was gained. Using non-ideal SMS models and considering their contact status among non-ideal surfaces, Zhang[23] modeled actual deformation with conjugate gradient and fast Fourier transform method, analyzing the deep coupling effect of geometric error and contacting deformation. It is mentioned for large-scale and thin-walled parts, due to assembly loads and initial geometric errors, forced clamping and joining actions are often adopted to eliminate clearance deviation. However, external and internal stresses will be generated, and it has a harmful effect on the service performance of assembled structures[24, 25]. For the above probability analysis of the EOCDCF event, the internal assembly stress function with practical error items should be further considered, and compensation strategies as deviation cannot meet design requirements are also urgent.

Then, to control assembly accuracy and avoid EOCDCF event in advance, by dividing error sensitivity into three indicator levels, Wu[26] proposed a quantitative analysis method with Monte Carlo simulation at features level, and its feasibility on multi-stations was also verified. However, this method cannot provide theoretical support for improving accuracy's retaining ability. It's known that non-rigid assembly simulation requires complex and difficult modeling tasks, to manage geometrical deviations during product's entire lifecycle, Polini[27] introduced a DT tool and established a continuous and unambiguous flow of variation from part design to final assembly. Where, available actual data sets from practical assembly sites were used to develop new and accurate simulation models, such as part's form deviations, fixture pressure, positioning errors, bonding amounts, assembly sequences, geometric shapes and parts' positions. Then to enhance the reliability of assembly quality, Yoshizato[28] stated that with accurate and timely adjustment strategies, unexpected assembly deformations and residual stress distributions could be predicted and reduced in advance. It is known that keeping geometric characteristics in a stable statistical state, is important to ensure the consistency among assembly error items of different structures. Guo[29] proposed optimization methods with ASFF (assembly station flowing fluctuation) analysis and feedback actions, adopting controlling measures based on historical data to reduce quality loss in aeronautical batch production was adopted. However, the SPC (statistical process control) solution is off-line and cannot realize real-time analysis. When assembling complex thin-walled structures, it is suggested that on-line feedback controlling mode needed to be promoted both for small batches and large scales, i.e., it's urgent to develop accuracy reliability analysis methods with strong applicability and good repeatability.

However, with above preventive actions, once the EOCDCF probability exceeds defined threshold, compensation operations on key parts have to be carried out. Although grinding and shimming operations would restrict the assembly efficiency in batch production, the repair compensation method is often adopted at the final stage of practical engineering site to control the geometric accuracy and internal stress[30]. This method refers to compensation on closed error links by repairing the sizes of certain error rings in assembly dimension chain. The comprised error links that to be repaired are defined as repair rings or repair areas, and the repair size is called as repair quantity. In practical engineering, to predict the detailed repair scheme for the workpieces, is mainly based on personal experience of technicians and workers, and there is no clear repair guidance. This situation often leads to trial assembly and repeated disassembly. Taking actual physical factors into account, for tolerance analysis or gap estimation during shimming process, Falgarone[31] presented AnatoleFlex software to create a complete and realistic assembly tolerance simulation that considers assembly sequence, joining defaults, composite material properties, form default, and contacting modeling. And the measured actual geometric deviation was taken as the input of simulation model. To minimize the repair workload, Heling[32] established tolerance-cost optimization model considers assembly dimension chain with statistical methods. Based on difficulty coefficient evaluation, Tlija[33] proposed tolerance allocation methods with Lagrange factor, which can evaluate assembly difficulty quantitatively and reduce repair probability. By investigating the effects of repair compensation on deviation propagation, Wang[34] presented a tolerance analysis method for composites assembly based on T-Maps method, focusing on accumulation of anisotropic variations and propagation of clamping force modification and shimming operations. Considering the realistic models with geometrical surface deviations are essential for further functional analysis, Anwer[35, 36] stated SMS models could represent parts' geometric deviations. Where, measurement data and actual models can be used to compensate for the accumulated error. In summary, to ensure the final dimensional error, current solutions for repair quantity are mainly based on the calculation of geometric dimension chain. The analysis result provides the repair quantity range on the part, and it is difficult to determine the specific repair area and value size in advance. As a result, this situation results in a lack of quantitative and scientific guidance in practical engineering, and it has a limit effect on the reliability improvement in assembly accuracy.

Aiming at EOCDCF event and extensive repair operations, novel solutions for improving accuracy reliability with probability analysis and prior precise repair compensation before practical assembly operations were studied. Section 6.2 considers

Digital Assembly Coordination and Quality Controlling Technology
for Aeronautical Thin-walled Structures

the internal stress function with practical multi-dimensional error items that having non-independent relationship, and probability model for accuracy reliability evaluation is modeled. Section 6.3 establishes the occurrence probability model of EOCDCF event with Gartner-Ellis theorem in LDP, and adopts the dynamic sliding window method to calculate its probability. Section 6.4 proposes the accurate repair simulation method based on IPSO algorithm with actual deviations, and designs detailed repair scheme in advance. Finally, wing-box component is taken to verify the effectiveness of improving accuracy reliability.

6.2 Probability Model of Assembly Error for Accuracy Reliability Evaluation

The multi-dimensional assembly error sources are complex and diverse, having the typical characteristics of non-independent and identical distribution, which would make it difficult to evaluate the reliability of assembly deviation accurately. In this section, to identify the sensitive error items and their accumulation effects in actual assembly, firstly, the modeling and coupling process of different variation sources that considering measured data and internal stress are analyzed. Then considering the causes of EOCDCF event in the detailed assembly procedure, a probabilistic evaluation model for assembly accuracy reliability is proposed.

6.2.1 Multi-dimensional Deviation Source Modeling Considering Actual Error and Internal Stress Effect

The out-of-tolerance phenomenon in practical assembly process, is demonstrated when the values of key geometric quality characteristics does not meet the design indexes. From a macro perspective, the assembly deviation is comprised of deviation items of external environmental and product itself. In fact, the final deviation can be mainly decoupled into the combination of initial manufacturing error, positioning error with assembly fixture, and deformation error. And deviations caused by measurement instruments, temperature and humidity changes, and transportation can often be ignored. It's known that deviation coupling based on geometric vector's superposition has a weak physical significance. Considering the measurement operations before and after the assembly process, the coupling and focusing effects of multi-dimensional actual error sources and their internal stress distribution with assembly loads should be concerned for accurate deviation modeling, and then, product state can be compared with theoretical model to obtain the deviation coupling result.

6.2.1.1　Initial Manufacturing Error with Model Reconstruction

To facilitate the accumulation calculation of assembly deviation, geometric error state should be measured and preliminarily analyzed before parts are located on assembly jig. For part's initial manufacturing error modeling and reverse reconstruction based on measurement operations, such as high precision scanning and KAF (key assembly features) are main focus for efficiency reasons. After geometric elements of KAF in design model are replaced with reconstructed shapes, stitching and materialization operations, i.e., virtual-reality fusing can generate the product reconstruction model that represents actual position and shape error distribution.

The reconstructing error is defined as deviation between reconstructed and the actual models, which includes feature's geometric error, measurement error, and reconstruction algorithm error. The measurement error is caused by the measuring instruments and manual operations, and the nominal accuracy of the instrument is often considered. The reconstruction algorithm error is generated by fitting the point cloud data with CAD modeling, and it can be obtained by the software's error detection function. To sum up, it can be expressed as $\varepsilon_{construction}=\sum(|\varepsilon_{measurement}|,|\varepsilon_{algorithm}|)$, and its deviation range is represented as $(-\varepsilon_{construction}\sim\varepsilon_{construction})$.

For qualified parts, the initial manufacturing error is considered as a constant value in the subsequent assembly process, so the stress effect caused by error factors under certain assembly conditions can be regarded as having a fixed distribution state and numerical size. Taking the assembly stress effect into account, it can be calculated with finite element method (FEM). However, in practical assembly process, it will lead to higher analysis complexity and calculation time. Previous studies have proved that superposition expression of Legendre polynomials with items 2 to 8, would conform to FEM analysis results considering the actual geometric error[37], and it can be adopted to characterize the assembly stress under force equilibrium state. Since the stress state has characteristics of force and moment balance, then the combination of Legendre polynomials is adopted to characterize the stress effect of initial manufacturing error.

6.2.1.2　Positioning Error of Assembly Fixture

At assembly stage, parts and components would subject to the locating and shape-retaining function of assembly fixture. For complex thin-walled parts, in the process of locating, posture adjustment, and clamping, firstly, the coordinate reference of fixture and assembly should be established, and then, parts are moved to the specified assembly station. Once positioning error meets the requirements, the fixture is locked in current locating status. After the entire procedure is completed, the assembled structure is removed from the fixture system. According to the interaction between workpieces and fixture, control function of locating positions and profiles shapes is divided into three stages: ① part locating and clamping (represented as S_1), ② position and shape adjustment (denoted as S_2), and ③ product removal from the fixture (denoted as S_3). For stage S_1, the fixture would contact parts, and plays a preliminary role in maintaining structure's shape. However, due to the initial manufacturing error, the stress-strain effort between fixture and parts would emerge, i.e., the contacting stress σ_T is introduced. At the end of this stage, there is an assembly geometric deviation K_{S_1} occurs, and the parts are located and clamped in place. From the second stage, the tooling plays the role of positioning and maintaining shape. For stage S_2, the assembled intermediate structure and fixture system can be regarded as a parallel mechanism, with the fixture's main function being considered as the detailed adjustment for reducing the out-of-tolerance phenomenon with forced clamping. Adjustment operations on the fixture are necessary for matching structure's actual posture and position with ideal model. With FEM analysis, the internal stress field can be taken as the differential value of the established displacement field. Therefore, the stress effect that generated by positioning error at the contacting areas, can be regarded as the differential value derived from the deformation displacement field of the tooling system. Given geometric state at the contacting interface of fixture and parts K_t, and clamping force that introduced by the forced assembly F_{S_2}, it is necessary to eliminate K_{S_1} by F_{S_2}.

From the above analysis, it's obvious that there is coupling relationship for positioning error in the above two stages. After forced clamping operations in stage S_2 is completed, the fixture is locked to execute the subsequent assembly procedures. Finally, after the joining and measurement operations is completed, the assembled structure is separated from the fixture. For stage S_3, there are no external assembly loads, and the internal stress generated in the previous assembly stages would be released, then the final deformation occurs.

In conclusion, the stress effect corresponding to positioning error at each stage can be represented as:

$$V_\sigma(T) = \begin{cases} \sigma_T, & S_1 \\ V_\sigma(K_t), & S_2 \end{cases}$$

$$V_\sigma(K_t) = g(F_{S_2}, K_t - K_{S_1}) \tag{6-1}$$

Where $V_\sigma(\cdot)$ represents the stress action corresponding to the geometric error state, and $g(\cdot)$ represents the stress action calculated by clamping force and the actual geo-metric state.

6.2.1.3 Deformation Error Caused by Assembly Loads

For typical thin-walled parts with a low rigidity, even if the fixture plays the shape-retaining function, the spring-back is difficult to predict due to the unstable working conditions. There is structure's deformation caused by immediate self-weight, and the passive rebound deformation due to clamping, drilling, and joining operations. The deformation deviation would lead to the complex stress actions and uneven distribution of internal stress for the complete component. For the stress effect with self-weight, the deformed geometric state ξ_G can be obtained by FEM simulation, and there is stress effect $V_\sigma(\xi_G)$ related to the actual geometric state. For the deformation deviation caused by riveting, screwing, and other joining processes, considering simulation results on stress field, it is still necessary to analyze deformation status after joining operations and the geometric shape error ξ_d. In conclusion, the entire stress effect can be represented as:

$$V_\sigma(\xi) = V_\sigma(\xi_G) + V_\sigma(\xi_d) \tag{6-2}$$

6.2.2 Probability Model Expression of Assembly Accuracy Reliability

Considering assembly deviation accumulation caused by multi-dimensional error sources and practical assembly conditions, and based on the tolerance analysis, this section integrates actual variation items into accuracy prediction with measured data. Specifically, the prediction basis is the part's reconstruction model, and the actual positioning and clamping information are also taken into account as modeling assembly deviation. With FEM solution, deformation error of flexible parts caused by gravity, clamping force, and joining force are simulated. The detailed accuracy prediction process, considering the above multiple error sources is shown in Figure 6-1. To be more specific, firstly, the part's reconstruction model with actual manufacturing error is coupled with fixture's positioning error. After the deformation due to the assembly loads is calculated with FEM analysis, it's superimposed on part's model. Subsequently, the assembly sequence and geometric constraint relationships are designed, and with the help of dimension analysis software, assembly dimension chain can be derived and updated. By updating the assembly deviation transfer matrix and using Monte Carlo algorithm, with the prediction results, the practical assembly site could be guided and adjusted iteratively.

However, the complex coupling relationships among above error sources would pose great challenges to the reliability of assembly accuracy. The influence of deviation coupling on accuracy's reliability can be explained by considering assembly historical data as an example. Taking assembly procedure or stage as the minimum action interval and the assembly unit, it can be considered that the transferred assembly deviation exist in

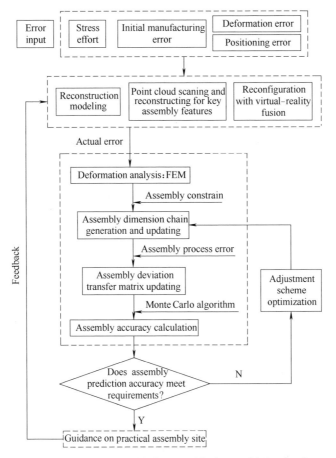

Figure 6-1   Assembly accuracy prediction considering multiple actual error sources

each assembly unit. As the assembly procedures progress, deviations that generated in each assembly unit are inherited and accumulated across different units. In detail, each assembly unit contains many assembly steps, such as positioning, drilling and joining. The generated positioning errors and contacting deformation error in these steps would be transmitted to next assembly unit, and their actual coupling relationships at different assembly stages can be shown in Figure 6-2. Where, the vertical axis stands for the size values of various deviation types, and each error source is collected from practical engineering site.

In Figure 6-2, it can be seen that except for the initial constant manufacturing error, other error types have a non-uniform distribution and a non-linear accumulation relationships. However, with uncontrollable assembly loads and working conditions, the deviation coupling has a focusing effect, and their accumulation might lead to a maximum out-of-tolerance phenomenon for some assembly units, preventing the assembly accuracy from being maintained in a stable state. They float up and down from the central deviation line, and there is no single upper or lower deviation type in the assembly process. As a result, the reliability of assembly accuracy would suffer great losses. From another perspective, with statistical analysis from engineering site, it's proved that assembly unit with the EOCDCF event often have quality problems, such as the deviation of skin profile, step difference, and mating gap, after the assembly work and even during assembly procedures or steps. It's mentioned that, only with reluctant concession, the unqualified assembled structure could be accepted. In addition, the EOCDCF phenomenon would also increase assembly costs. Therefore, repair operations such as shimming and grinding, should be adopted to meet geometric assembly requirements. However, the compensation actions would make it difficult to meet technical indicators, such as the success rate of one-time assembly.

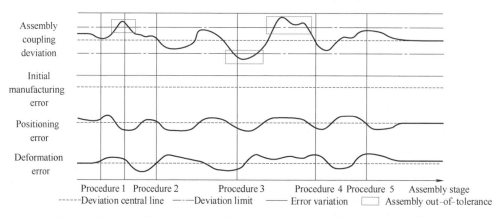

Figure 6-2 Assembly error generating and coupling at different assembly stages
(See the color illustration)

For the above assembly deviation generation and coupling effects, it indicates that even if the assembly processes are under the controlled state, the accumulated deviation could still be located at the edge of out-of-tolerance line, i.e., the reliability of assembly accuracy are severely challenged. The deviation forming diagram shown in Figure 6-2 has intuitive characteristics, and it is conducive to analyze the accuracy's reliability directly and could also provide beneficial guidance for subsequent assembly. Therefore, in order to keep assembly accuracy in a stable and controllable status, intuitive and user-friendly accuracy reliability evaluation solutions should be adopted. Based on the stress effect caused by actual multi-dimensional error sources and the intuitive characteristics of the probability method, the deviation coupling characterization can be represented by Equation (6-3), and the reliability evaluation probability model for assembly accuracy is established with Equation (6-4).

Digital Assembly Coordination and Quality Controlling Technology
for Aeronautical Thin-walled Structures

$$\zeta(k) = f[V_\sigma(M), V_\sigma(T), V_\sigma(\xi)] \tag{6-3}$$

$$P_{EVCFCO} = P[\zeta(k) > \zeta_m] \tag{6-4}$$

Where $P(\cdot)$ refers to the occurrence probability of out-of-tolerance event; $\zeta(k)$ represents the deviation coupling and focusing function; and ζ_m stands for the deviation coupling effect during the out-of-tolerance situation. The EOCDCF phenomenon is considered to have occurred once $\zeta(k)$ exceeds ζ_m. For P_{EOCDCF} in the specific assembly procedures and steps, it can be obtained through historical and online data. The above model establishes a preliminary judgment method for reliability damage on assembly accuracy. However, the model parameters would have a great change with a minor adjustment on assembly process. Combining with actual assembly process, the above probability model is difficult to solve.

6.3 Probability Analysis and Precise Adjustment of Out-of-tolerance Event with LDP Method

Taking the physical meaning of probability model into account, it can be seen that the key to reduce the reliability damage of accuracy is predicting the maximum peak (denoted with the box in Figure 6-2), i.e., the early perception of quality problems. Considering the actual measured error data in each assembly procedure and step, this section starts with the prediction of maximum out-of-tolerance event. Then, by estimating the probability of EOCDCF phenomenon with LDP, the online probability calculation model with dynamic sliding window method can guide the practical assembly process precisely and in real-time.

6.3.1 Probabilistic Model for Reliability Evaluation on Accuracy with LDP

From the perspective of actual assembly information and conditions, and based on the deviation relationship between the net increase in the micro time slots and the allowable tolerance, probability problem modeling of out-of-tolerance event for a certain assembly process is deduced, and then, the probability of EOCDCF event is estimated with the Gartner-Ellis theorem in LDP.

6.3.1.1 Probabilistic Problem Modeling Considering Actual Assembly Process

Assuming that an assembly procedure with a duration time τ is carried out at a normal speed according to the designed assembly period, and key quality characteristics could be obtained in real-time at each micro time slots w_i, the net increase of the assembly deviation is expressed as follows:

$$\Delta\zeta_w = \zeta_{wj}^C - \zeta_{wj}^H \tag{6-5}$$

Where w is determined by the data acquisition frequency in each assembly procedure, and it is often consistent with the minimum assembly procedure or steps. With w, it is convenient for process variation modeling and rapid adjustment for the subsequent assembly process; ζ_{wj}^C represents the deviation increase during one time slot w; and ζ_{wj}^H stands for the deviation compensation during the time slot w. Then for the current probability of the net increase, it can be expressed as $\pi_h = P(\Delta\zeta_w = h)$. If $\Delta\zeta_w > 0$, it can be taken as an increase in deviation during w, and vice versa.

According to the results of assembly tolerance allocation, for the basic assembly procedures of flexible and thin-walled parts, i.e., "positioning-clamping-drilling-joining-releasing", the maximum allowable deviation range ζ_τ^n of each assembly procedure is given, where n represents the current procedure or stage. Therefore, after the operations within $w \to w+N$ time slot, the superposition of deviation's net growth in the above N micro time slot can be represented as:

$$\Delta\zeta^{w+N} = \sum_{i=0}^{N} \Delta\zeta_{w+i} \tag{6-6}$$

By calculating the difference between the net growth and the maximum allowable deviation of the assembly process, then the remaining allowable deviation range during the N time slots can be represented as:

$$\zeta_j^{w+N} = \zeta_\tau^n - \Delta\zeta^{w+N} \tag{6-7}$$

If the calculation result is less than zero, it can be considered that EOCDCF event occurs. Combining with Equation (6-4), the occurrence probability of EOCDCF event can be expressed as:

$$P_j^{w+N} = P(\zeta_j^{w+N} < 0) = P(\zeta_\tau^n < \Delta\zeta^{w+N}) \tag{6-8}$$

Furthermore, considering the statistical parameters that obtained during the above N multiple slots, Equation (6-8) can be updated by combining the acceptable deviation increase and their actual variation status. Within N time slots, the acceptable average deviation increase can be defined as:

$$b_j = (\zeta_\tau^n - \zeta_{\tau min}) / N \tag{6-9}$$

Where, $\zeta_{\tau min}$ represents the minimum allocation tolerance when the historical quality data or simulation data have an out-of-tolerance phenomenon.

And the expectation of deviation variation in N time slots during actual assembly process is defined as:

$$g_j = E\left[(\sum_i^N \Delta \zeta_{w+i}) / N \right] \tag{6-10}$$

As a result, when the deviation range is greater than the acceptable deviation increase, i.e., $g_j > b_j$, the progresses of the assembly operations might lead to the out-of-tolerance phenomenon, and the occurrence probability of EOCDCF event is represented as $P_j^{w+N} = P(g_j > b_j)$. Correspondingly, then the updated Equation (6-8) can be rewritten as:

$$P_j^{w+N} = P\left[(\sum_i^N \Delta \zeta_{w+i}) / N > b_j \right] \tag{6-11}$$

It is mentioned that, for the probability model of Equation (6-4), it is improved with the consideration of actual assembly process and real-time error status, establishing the relationship between probability modeling and actual assembly process. However, for the general expression of probability model in Equation (6-11), without considering the actual assembly conditions and the coupling relationships of different error sources, it's difficult to calculate the occurrence probability of EOCDCF event. To be more specific, to enhance its solvability, Equation (6-11) should be decomposed into more detailed solutions and methods. As a result, the combination of deviation analysis and LDP is considered for improving the occurrence probability modeling in the following section.

6.3.1.2 Probability Modeling Based on Gartner-Ellis Theorem in LDP

The mathematical description of LDP can be described as:

For a random error variable sequence $\{X_k\}$, where $k=1, 2,\cdots, n$, if there is a rate function I in the real number set $\mathbf{R}$ such that:

① For any closed set $F \subset \mathbf{R}$, there is $\limsup_{n \to \infty} \frac{1}{n} \log P(x_n \in F) \leqslant -\inf_{x \in F} I(x)$.

② For any open set $G \subset \mathbf{R}$, there is $\liminf_{n \to \infty} \frac{1}{n} \log P(x_n \in G) \geqslant -\inf_{x \in G} I(x)$.

Where $P(\cdot)$ refers to the occurrence probability of an EOCDCF event. Then the entire random sequence $\{x_n\}$ could be considered to satisfy LDP with the rate function I.

In engineering applications, the most considered situation is to adopt Cramer theorem[38] to analyze error variables with independent and identical distribution. However, the above condition neglects the non-linear coupling effect among various error items. In fact, it's an ideal simplification of practical engineering problems, which would lead to an obvious difference compared to actual assembly data. To be more specific, for the complex coupling relationships among manufacturing errors, positioning errors and deformation errors, it is necessary to analyze the above non-independent and multi-dimensional variation sources with the Gartner-Ellis theorem[39]. Taking the universal applicability of the conclusion derived from Cramer theorem into account, the Gartner-Ellis theorem in LDP can be briefly expressed as follows.

To $\forall \lambda \in \mathbf{R}^d$, given the logarithmic generation function that describes its convergence property is defined as $\Lambda(\lambda) \triangleq \lim\limits_{n \to \infty} \dfrac{1}{n} \Lambda_n(n\lambda)$, and the definition domain of its variables is $D_\Lambda \triangleq \{\lambda \in \mathbf{R}^d : \Lambda(\lambda) < \infty\}$. For any closed set $F \subset \mathbf{R}^d$, there is $\limsup\limits_{n \to \infty} \dfrac{1}{n} \log \mu_n(F) \leqslant -\inf\limits_{x \in F} \Lambda^*(x)$, Where μ_n represents the probability distribution of the random error variable x, $\Lambda^*(x)$ represents the rate function I that obtained after $\Lambda(\cdot)$ has a Fenchel-Legendre transformation, donated as $\Lambda^*(x) \triangleq \sup\limits_{\lambda \in \mathbf{R}^d} [\lambda x - \Lambda(\lambda)]$.

According to the Gartner-Ellis theorem, for the deviation variables in the coupling result $\zeta(k)$ considering multi-type error items, which are non-independent and identical distributed, the extreme accumulation deviation, i.e., the occurrence rarity of EOCDCF event, could satisfy the LDP. In order to avoid its occurrence at the kth time nodes, the distribution range of $\zeta(k)$ should be located within the value range of $(0,[a])$. Where $[a]$ represents the upper limit of allowable tolerance during assembly process, corresponding to the maximum allowable deviation ζ_τ^n of each assembly procedure, then there is:

$$\limsup\limits_{N \to \infty} \frac{1}{N} \log P \left[(\sum\limits_i^N \Delta\zeta_{w+i}) / N > b_j \right] \leqslant -I(b_j) \tag{6-12}$$

Where $I(b_j) = \sup\limits_{\lambda \geqslant 0} \{\lambda b_j - \log M(\lambda)\}$ is a convex rate function, $M(\lambda)$ represents the moment generating function, where $M(\lambda) = \sum\limits_{h=0}^{[a]} \pi_h e^{i\lambda}$. π_h stands for the probability distribution related to $\Delta\zeta_w$. For the inequality expression of probability calculation result $\zeta(k)$ that derived by Gartner-Ellis theorem, without losing the modeling generality, if N is large enough, then the occurrence probability of EOCDCF event can be approximately calculated as:

$$P_{EOCDCF} \approx \{-NI(b_j)\} \tag{6-13}$$

Furthermore, based on Equation (6-12) and (6-13) which combining deviation analysis and LDP method, the probabilistic model for reliability evaluation on assembly accuracy has been established. For this model, it would transform the original out-of-tolerance judgment based on historical deviation data into the out-of-tolerance

Digital Assembly Coordination and Quality Controlling Technology for Aeronautical Thin-walled Structures

probability analysis based on actual measured variation items. Compared to the model in Equation (6-4), the ability of solving the problem with actual assembly data and online error information could be increased. After the probability model is established, its solution method is then to be studied.

6.3.2 Solution and Analysis for Probability Model Based on Dynamic Sliding Window Method

In order to solve the probability model that established in Equation (6-12) and (6-13), it is necessary to obtain the acceptable average deviation growth b_j, the probability distribution π_h, and expected value g_j of deviation's variation range. The value of b_j can be obtained according to engineering experience and the existing assembly data, and the values of g_j and π_h are related to the distribution principle of $\Delta\zeta_w$. However, $\Delta\zeta_w$ is generally considered to have a normal distribution during the assembly deviation analysis, then this hypothesis cannot reflect the actual assembly situation. From another aspect, the contribution of different assembly procedures and product's key characteristics to assembly quality is significantly different, and their difference should be highlighted. In addition, considering the assembly work locates at the end of the entire manufacturing process, the remaining period for assembly is always tight, then the assembly efficiency could be significantly reduced due to the excessive process adjustment and manual repairing work. Aiming at this situation, according to the calculated probability information, appropriate and suitable process adjustment methods should be adopted.

In this section, firstly, solving method for the probability model with the dynamic sliding window solution is proposed, allowing the probability distribution of $\Delta\zeta_w$ with the measured deviation items to be characterized. Then, reliability adjustment strategies for assembly accuracy assurance are proposed according to the solution results of above probability model.

6.3.2.1 Probability Calculation with Sliding Window and Measured Data

The sliding window method[40] could estimate the probability distribution according to the occurrence frequency of different results. To be more specific, for its applicability with assembly technology, by analyzing the acquisition frequency and allowable tolerance of the quality data during each assembly step, the micro interval can be divided firstly. Then, for the quality data, its probability distribution can be converted as the ratio between the occurrence interval number and the total number. Based on the above analysis, the probability distribution estimation based on the actual information can be realized. And the sliding window method is adopted to estimate the probability distribution of $\Delta\zeta_w$.

Supposing the sliding window is represented as N_d, and by discretizing the value range of $\Delta\zeta_w$, i.e., $(0,[a])$ and dividing it into W segments, such as $\left(0,\dfrac{1}{W}[a]\right)$, $\left(\dfrac{1}{W}[a],\dfrac{2}{W}[a]\right),\cdots,\left(\dfrac{W-1}{W}[a],[a]\right)$, then the estimated results of π_h and g_j can be transformed into the estimation of occurrence number N_l for the events $\Delta\zeta_w = l$ in the sliding window. Where $\Delta\zeta_w = l(1 \leqslant l \leqslant W)$ represents the value $\Delta\zeta_w$ located at the lth segment. And the obtained preliminary estimation can be shown as:

$$\eta_f(n) = N_l / N_d \qquad (6\text{-}14)$$

Considering the above result is highly sensitive to the value of N_d and N_l, then the forgetting factor (or smoothing factor) ρ is introduced here to make the obtained probability distribution smoother:

$$\hat{\pi}_h(n) = \rho\hat{\pi}_{h-1} + (1-\rho)\eta_f(n) \qquad (6\text{-}15)$$

Where the forgetting factor has a distribution range of $\rho \in [0,1]$, and its value determines the weight ratio between historical estimation $\hat{\pi}_{h-1}$ and current estimation $\eta_f(n)$. To reduce the impact of abnormal data noise for the current estimation, relevant studies suggest the value range of ρ should be taken as [0.7, 0.9] [19]. Then by substituting it into Equation (6-15), the estimated value π_h can be obtained, and the estimated g_j can be represented as:

$$g_j = \left(\sum_{i=W-N_l+1}^{W}\Delta\zeta_{wi}\right) / N_d \qquad (6\text{-}16)$$

For the above description on sliding window with a fixed interval size, it can calculate the occurrence probability of the EOCDCF event in a certain time series initially. However, for the generation and transmission of actual process data items, they may change randomly as assembly procedures progress. From the perspective of detailed assembly operations, the measurement and inspection frequency should be increased appropriately for key processes that have a great contribution to final quality index. From another perspective, for the measurement of key product characteristic, the data acquisition density should also be assessed per unit area. However, for the designed micro gap w, it is not economical to increase the detection frequency and density during the whole assembly process. As a result, according to the actual assembly process, solutions

Digital Assembly Coordination and Quality Controlling Technology
for Aeronautical Thin-walled Structures

such as: ① dynamical adjustment of the interval size of sliding window, and ②
number increasment of micro gaps for the key assembly procedures, product
characteristics, and control characteristics, can focus a special attention on the assembly
process and geometric features that have a significant impact on product's overall
performance and assembly accuracy.

When there is a micro slot that gets close to the target value l in the original sliding
window, for the dynamic sliding window method, the adjacent micro slot would be
divided by several times. Then for the certain section N_j in N_d that adopted to
enhance the observation frequency and density, it is to be divided into n sections, where
the occurrence number of $\Delta \zeta_w = l$ is displayed as N_{lj}, as shown in Figure 6-3.

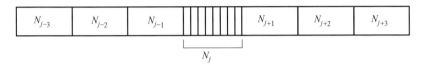

Figure 6-3　Schematic diagram of dynamic sliding window method

Then for the entire assembly process, after the dynamic sliding window is
constructed k times, the Equation (6-14) that can be modified as:

$$\eta_f(n) = (N_l + \sum_{j=m}^{m+k} N_{lj}) / (N_d + \sum_{j=m}^{m+k} N_j) \qquad (6\text{-}17)$$

With the above modeling steps, the dynamic sliding window method can be adopted
to solve the probability model of assembly accuracy. Considering the fact that this
method takes the measured actual assembly information as the input data, as a result, in
practical assembly process, the calculated online probability results can be used to
evaluate whether the accuracy reliability is suffering from damage.

6.3.2.2　Assembly Accuracy Reliability Adjustment Strategy Based on Online Probability Calculation

With the aim of reducing the occurrence of EOCDCF phenomenon, combining the
above probability model for reliability evaluation with assembly accuracy based on
deviation analysis and LDP method, the corresponding reliability adjustment strategy is to
be explored in this section. With the online probability calculation results, the detailed
technical route is shown in Figure 6-4.

For the acceptable average deviation growth b_j that calculated for a time slot in
one certain assembly procedure, and the corresponding deviation's expectation g_j, when
$g_j \geqslant b_j$, it indicates the statistical expectation of the deviation's variation is greater than

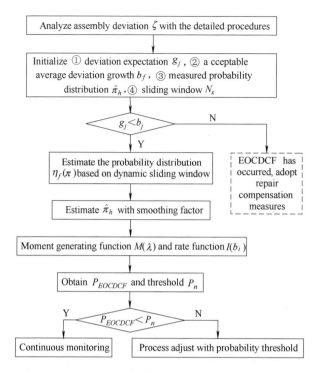

Figure 6-4 Technical route for accuracy reliability adjustment with online probability calculation

or equal to the acceptable average deviation growth. This situation is generally considered that EOCDCF phenomenon has occurred, and the reliability of assembly accuracy has been damaged. In this unstable status, it is necessary to take timely technical measures on assembly process to compensate assembly quality, such as grinding, shimming, or other optimal actions. If $g_j < b_j$, i.e., the statistical expectation is smaller than the acceptable average deviation growth. However, according to Figure 6-2, it cannot be assumed that EOCDCF phenomenon wouldn't occur, and the reliability of assembly accuracy is still to be tested. According to reliability analysis and modeling in above sections, the occurrence of EOCDCF phenomenon under this situation can be taken as the rare event. As a result, the corresponding probability of P_{EOCDCF} can be calculated by the above methods with Equation (6-3)~(6-17), as well as the technical route shown in Figure 6-4.

From the perspective of improving the accuracy's reliability, after obtaining the occurrence probability of EOCDCF event for a certain time sequence, the first thing to do is to judge the size relationship between the calculated P_{EOCDCF} and the threshold value P_n. Where P_n represents the probability threshold corresponding to the detailed assembly procedure n. For the situation of $P_{EOCDCF} < P_n$, the occurrence probability of

Digital Assembly Coordination and Quality Controlling Technology
for Aeronautical Thin-walled Structures

EOCDCF event in the current assembly step has not reached the threshold. To keep the assembly accuracy within an acceptable range, the monitoring on assembly quality should be carried out continuously. When $P_{EOCDCF} \geqslant P_n$, the occurrence probability of EOCDCF event has reached and exceeded the threshold value, and there is a low accuracy reliability. Under this situation, process adjustment measures should be taken to limit P_{EOCDCF} within the threshold. Combining with the current feedback adjustment solutions, the damage degree of accuracy reliability can be analyzed according to the difference value size between P_{EOCDCF} and threshold, and then the procedure adjustment mode can be classified through the above difference value. Since the threshold value P_n is usually selected between 0.5 and 0.7, there are four types of process adjustment mode, and the detailed adjustment actions in practical assembly engineering are summarized in Table 6-1.

Table 6-1   Process adjustment mode based on probability threshold relationship

Mode	Probability-threshold relationship	Damage degree	Process adjustment actions
1	$P_n < P_{EOCDCF} \leqslant 0.7$	Small possibility	Recheck the position and shape of products with assembly tooling adjustment
2	$0.7 < P_{EOCDCF} \leqslant 0.8$	High possibility	Improve the measurement and sensing frequency and density for assembly information, expand the selection range of dynamic sliding window, and repeat the adjustment actions in Mode 1
3	$0.8 < P_{EOCDCF} \leqslant 0.9$	Great possibility	Repeat the adjustment method in Mode 2, select the assembly operation parameters with more robustness (such as a lower drilling feed rate), adjust the removal method of assembled structure to give priority for the areas with variable curvature and opening gap, and optimize the tolerance allocation of key components such as skin profile and hinge joints
4	$0.9 < P_{EOCDCF} \leqslant 1.0$	Damage is to be happened	Repeat the adjustment method in Mode 3, stop the assembly work to check and ensure the deviation source information sensing method is appropriate, recheck the detailed assembly parameters

6.4   Precise Optimization of Repair Quantity Based on Measured Data and IPSO Algorithm

As demonstrated in Table 6-1 and the dotted box in Figure 6-4, for the occurrence of out-of-tolerance event, i.e., $P_{EOCDCF} > 1$, it indicates an unstable assembly process, with the reliability of assembly accuracy falling to meet the requirements. Under this situation, the assembly structures need to be accurately repaired and compensated. This solution is especially important to ensure final geometric accuracy and control structure's internal stress. However, on the premise of ensuring the final geometric accuracy, the key aspect of improving accuracy reliability is to determine the part's specific repair area and quantity accurately in advance. Then the quantitative and scientific guidance on engineering site could be designed. Basically speaking, the repair quantity refers to the amount of shimming or grinding operations on the mating and coordination areas among assembly parts. For large thin-walled structures, considering the cumulative accumulation effect of different error sources, there is a complex nonlinear relationship between the final compensated geometric accuracy and the repair amount of each key focus area. Then the mathematical model of repair quantity optimization can be expressed as:

$$\mathrm{Min}E(x) = f[(x_1 \pm \varepsilon),(x_2 \pm \varepsilon),(x_3 \pm \varepsilon),\cdots,(x_n \pm \varepsilon)] \qquad (6\text{-}18)$$

Where $E(x)$ represents the target assembly deviation after repair compensation; x_i stands for the repair amount of the ith repair area or the ith error link, Where a positive value indicates shimming operation, while a negative value indicates polishing or grinding operation; ε is a small quantity represents the repairing error, which is related to the shim's manufacturing error and the specific repair process, and it has a range of $0 < \varepsilon < |x_i|$; Then, $(x_i \pm \varepsilon)$ would stand for the repair range of the ith repair area. According to engineering experience, the repair range should not exceed 1mm under normal assembly conditions, namely, $-1\mathrm{mm} \leqslant (x_i \pm \varepsilon) \leqslant 1\mathrm{mm}$.

To realize the precise design of repair quantity, two optimization processes are proposed. Firstly, considering the detailed "assembly error coupling and probability model for accuracy reliability evaluation" in Section 6.2, and the technical routine "assembly accuracy prediction considering multiple actual error sources" in Figure 6-1, a group of repair schemes that satisfy final assembly quality requirements would be gained. From the perspective of optimization solution, for the optimization problems shown in Equation (6-18), they can be taken as local optimal results. This may not help technicians to achieve the optimal assembly decision-making efficiently. Then considering PSO (particle swarm optimization) algorithm has advantages of not relying on the strict mathematical properties of the optimization objective, it is easy to be realized in engineering. In addition, it is also a non-deterministic optimization algorithm that have more opportunities to search for the global optimal solution. As a result, it is very applicable for the repair quantity optimization calculation across different key repair areas. However, for the repair scheme that considering assembly accuracy and internal stress, to avoid the optimization solution from falling into the local optimum and improve its calculation efficiency, this section proposes an IPSO optimization method for repair quantity design at different key areas[41]. As a result, the second optimization process is to find the best solution among the repair schemes in the previous step, and their relationship can be shown in Figure 6-5. To be more specific, by combining repair

Digital Assembly Coordination and Quality Controlling Technology
for Aeronautical Thin-walled Structures

simulation and accuracy prediction, a group of current optimal repair quantities for each region is taken as the input of IPSO calculation, and repair simulation is carried out based on this group of repair schemes to obtain the target assembly accuracy. Where, the deviation value is substituted to IPSO algorithm as the fitness function, so as to participate in the subsequent iterative calculation. After the termination condition is satisfied, the iteration is stopped, and repair quantities of each repair area with the above repetitive optimization process can be obtained, resulting in the final optimum repair scheme.

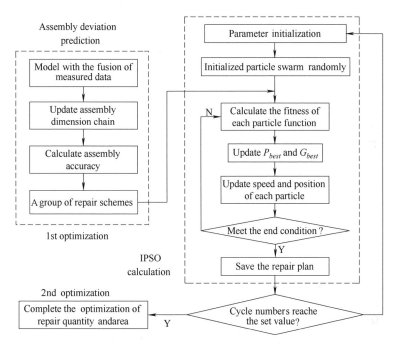

Figure 6-5   Twice optimization calculation of accurate repair quantity based on IPSO algorithm

For the IPSO calculation, the random weight and learning factor with asynchronous changing are integrated into standard PSO, then the algorithm has a stronger searching ability on repair quantity's design. The iterative formula of particle's velocity and position of IPSO can be described as:

$$v_{i,t+1} = wv_{i,t} + c_1 r_1 (pbest_i - x_{i,t}) + c_2 r_2 (gbest - x_{i,t}) \tag{6-19}$$

$$x_{i,t+1} = x_{i,t} + v_{i,t+1} \tag{6-20}$$

Where $x_{i,t}$ and $v_{i,t}$ represent the position and velocity of the ith particle at time t, respectively; w stands for the random inertia weight; c_1 and c_2 represent the asynchronous learning factors; r_1 and r_2 stand for the random factors (with numerical range from 0 to 1); $pbest_i$ represents the current optimal individual of the ith particle; and $gbest$ stands for the current global optimal individual. It is mentioned that, for the random weight W, it can be expressed as:

$$W = \mu_{min} + rand() \times (\mu_{max} - \mu_{min}) + \sigma \times randn() \tag{6-21}$$

Where μ_{min} and μ_{max} represent the minimum and maximum value of random inertia weights, respectively; $rand()$ stands for a random number that uniformly distributed in [0, 1]; $randn()$ stands for a random number with a normal distribution; and σ represents the standard deviation.

For the learning factors c_1 and c_2, they can be expressed as:

$$c_1 = (c_{1e} - c_{1s})(t - t_{max}) / t_{max} + c_{1e} \tag{6-22}$$

$$c_2 = (c_{2e} - c_{2s})(t_{max} - t) / t_{max} + c_{2e} \tag{6-23}$$

Where c_{1s} and c_{2s} represent the starting values of the factors c_1 and c_2; while c_{1e} and c_{2e} stand for their termination values, under normal conditions, $c_{1s}=c_{2s}=0.5$, $c_{1e}=c_{2e}=2.5$, with c_1 and c_2 increasing and decreasing asynchronously.

When IPSO algorithm is adopted for the optimization calculation on repair quantity, the following relevant parameters should be defined according to the specific repair optimization model and repair schemes and operations.

① Particle's dimension represents the number of repair areas. In this chapter, the dimension d of particles is set to be less than 4 according to calculation experience and practical assembly objects.

② For the position matrix $x_{i,t}$ of the particle, it represents the gained repair scheme that is suitable for the current particle, which can be updated by Equation (6-20). Each numerical value in matrix $x_{i,t}$ represents the repair amount that distributed on each key area. Since the repair amount should not exceed 1mm, there is $|x_{i,t}(j,1)| \leqslant 1$.

③ For the particles' velocity matrix $v_{i,t}$, it determines the search direction and step size of the position matrix and can be updated by Equation (6-19). To be more specific, it represents the increment amount of the repair quantity optimization. In order to avoid the rapid growth of particle's position, there is a limitation of $|v_{i,t}(j,1)| < 0.2$, because it would lead to the decrease of optimization ability.

④ For the fitness of PSO algorithm, it is the judgment index for the quality evaluation of each particle. For the position deviation of the curved surface, the fitness value stands for the variation after the repair compensation measure has been applied. The calculation basis is the surface's position deviation, which can be obtained with the repair simulation results in virtual environment.

⑤ For the typical repair area with curved surface on complex and thin-walled structures, the termination condition of IPSO optimization is set such that the repaired surface's position deviation could meet design requirements, that is to say, the optimized repair scheme can make the surface's assembly deviation qualified.

By defining the above parameter values and repair operation constraints, and combining the optimized fitness values obtained according to repair simulation, a group of feasible repair schemes could be gained. Considering IPSO is a non-deterministic optimization algorithm, the results obtained in each optimization step are different, because there are multiple groups of local optimal solutions. Then by repeating the above two optimization processes, the final optimal repair scheme can be obtained. As a result, reliability of assembly accuracy can be improved with the guidance on the repairing operations at practical engineering site.

6.5 SPC for Aeronautical Assembly Quality Control for Single Component

6.5.1 Overall Framework for SPC Technology

For aircraft assembly, the SPC technology that used for product key features originates from the advanced quality system (AQS) in Boeing. Its essence is to control the final assembly quality of the product in the manufacturing process. With this method, beneficial results can be gained, including ① the fluctuations of the product key characteristics can be reduced to a lower level, ② the statistical control states of the above characteristics can be kept, ③ the occurrence of assembly errors inconsistency can be minimized. For SPC, by monitoring and analyzing the fluctuation of quality samples that collected during assembly process, when the acquired actual value goes beyond the control range, it is necessary to find out the specific causes of assembly problems efficiently, such as whether the assembly precision or assembly stress meets the design requirements. Where, the assembly stress is defined as the internal residual stresses in the final assembled structure, which are caused by manufacturing errors of parts, positioning error of assembly jigs, clamping/drilling/jointing forces during assembly process. Then, relevant solutions would be determined, and with the adjustments on the specified manufacturing process, the purpose of assembly quality control can be achieved.

6.5.2 Analysis on SPC Chart in Aircraft Assembly with Qualitative Practical Experience

As is well known, the manufacturing/assembly for aircraft has the characteristic of medium to small scale production. When the production process is in a stable state, for the measuring and detecting accuracy data of quality samples, they are considered to obey the normal statistical distribution in most cases. It is assumed that for the two corresponding important parameters, i.e., the mean value and the standard deviation, they are independent of each other based on the statistical analysis of a large number of samples. However, the collected data of several measuring points for given aircraft components fall into the category of small samples in practical production statistics. For aviation products, even in the batch production phase, the sample size could not be particularly large. This would cause an insufficient information. Then, the distribution principle and the statistical parameters of the random variables are always unknown. As a result, the calculations that fitting for probability distribution for large sample data, such as parametric estimation, hypothesis testing, and regression analysis, also cannot be effectively applied to the small samples analysis. As a result, there is not enough data information to judge the assembly precision accurately, which would lead to a difficulty in the SPC analysis process.

For the purpose of expanding the statistical information of the measurement data maximally, and revealing the potential error accumulation principles, the construction process of the quality sample size is analyzed firstly. In the aircraft assembly process, the measurement characteristics, such as the measuring points corresponding to the final assembly error links, are usually distributed symmetrical for aircraft components. And with regard to the similar type or the similar assembly cases of the detection geometric features, if the manufacturing specifications or conditions, such as ① assembly quality require-ments (e.g., the deviation of skin profile), ② skills of workers and technicians, ③ process equipment/tooling, ④ measurement means, and ⑤ external conditions (e.g., the environmental factor), are basically the same as each other, then these characteristics can be considered to have the same fluctuation situation. Correspondingly, the above characteristics obey the same normal distribution, meaning the values of statistical mean

$(\bar{X} = \left(\sum_{i=1}^{n} X_i\right) / n)$ and standard deviation $(S = \left[\sum_{i=1}^{n}(X_i - \bar{X})^2 / (n-1)\right]^{1/2})$ are equal to each other.

With the above analysis, for the sake of solving the situation of insufficient data sample size, the data of these detection points can be combined together to construct a relatively larger amount of data samples which is helpful. It is noteworthy that only product quality data is combined here. At the same time, the standard transformation method can be utilized to transform the data that have a distribution of $N(0,1)$ from $N(\mu_i, \sigma_i)$. And then, the specific relevant analysis can be carried out in the statistical control chart. The chart adopts the conventional W. Shewhart control chart for quality control, and it is composed of central line (CL), upper control line (UCL) and lower control line (LCL), as shown in Figure 6-6. Where the x axis stands for the data samples, and the y axis represents the values of their

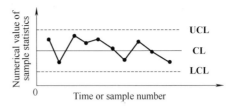

Figure 6-6　The SPC chart

quality characteristics. It is notable that the UCL and LCL are determined by the customer or the assembly requirements, which cannot be changed randomly as the manufacturing process progresses. And the range of UCL/LCL is often tighter than the tolerances required by customers or from design/manufacturing/assembly.

According to the detection values of key product elements in a specified manufacturing link or error link, the individual moving range chart, is used for statistical control in aircraft assembly process. X stands for the detection value of each quality sample. The moving range R_s refers to the difference value between different adjacent data items, and the sample number of 1. To the aircraft assembly process or operations, the construction principle of $X - R_s$ can be described as follows.

Step 1　Collect the quality sample data for m components, such as the positioning error of the tooling, the manufacturing precision of the comprised parts, the assembly accuracy or the coordination accuracy of the key assembly features. It is notable that these data is the foundation for SPC analysis.

Step 2　For each component, n sample measurement points are contained, and the values are denoted as: $x_{11}, x_{12}, \cdots, x_{1n}, \cdots, x_{i1}, x_{i2}, \cdots, x_{in}, \cdots, x_{m1}, x_{m2}, \cdots, x_{mn}$.

Step 3　Assuming the manufacturing and assembly process is in a stable status, which means each value has a normal distribution of $x_{ij} \sim N(\mu, \sigma^2)$, then for a certain component, the moving range R_s can be calculated by Equation (6-24).

$$R_{sij} = | x_{ij} - x_{i(j-1)} |$$ (6-24)

And the average R_s can be calculated as:

$$\overline{R_s} = \frac{1}{m(n-1)} \sum_{i=1}^{m} \sum_{j=2}^{n} R_{sij}$$ (6-25)

Step 4　For the individual control chart of quality sample[29], there is:

$$\begin{cases} UCL = \overline{x_{ij}} + 3\sigma = \overline{x_{ij}} + 2.66\overline{R_s} \\ CL = \overline{x_{ij}} \\ LCL = \overline{x_{ij}} - 3\sigma = \overline{x_{ij}} - 2.66\overline{R_s} \end{cases}$$ (6-26)

Where σ is the overall standard deviation of the data samples, which has a specific relationship with C.

For the moving range R_s control chart of quality sample, there is:

$$\begin{cases} UCL_{R_s} = \overline{R_s} + 3\sigma_{R_s} = 3.27\overline{R_s} \\ CL_{R_s} = \overline{R_s} \\ LCL_{R_s} = \overline{R_s} - 3\sigma_{R_s} = 0 \end{cases} \tag{6-27}$$

Where σ_{R_s} represents the standard deviation of R_s. The result of LCL_{R_s} is generally is less than 0, considering the definition of R_s, this result can be taken as 0, for the convenient analyses on the SPC datagraphic.

With the view of influencing factors and statistical analysis, some intuitive conclusions can be drawn regarding assembly quality control for the manufacturing system quickly, with the help of estimating the mean value and the standard deviation with the $X - R_s$ chart. If the measured sample points that described in the control chart exceed the control limits, it is considered that the statistical process is abnormal, and there are problems in the manufacturing process that lead to the phenomenon of out-of-bounds. With the analysis of the above chart, it is necessary to analyze the distribution status or trends of sample data in SPC chart. Actually, there are two main types-of-abnormal phenomenon: the sample measurement points exceed the limits, and the sample points are not randomly arranged within the boundary. On the basis of a large amounts of practice working, the criteria for judging the rules of SPC control chart have been summarized, and they have a mapping relationship.

According to the actual measured characteristics of aircraft assembly coordination process and the common problems occurred for the inconsistencies of different assembly errors, the specific rules for identifying the causes of the abnormal fluctuation that leading to the assembly accuracy problems, can be summarized as below. And the following criteria are mainly based on engineering experience and the company's regulations on quality management.

① When the position of some measurement points is just located at the upper and lower control lines, or located beyond the upper and lower control lines, it can be judged that there is a significant system error in the manufacturing and assembly process. Most of the time, the above phenomena are caused by the errors of measurement system, the significant changes of assembly environment, the instability in automatic controls of FAT (flexible assembly tooling), the obvious shortcomings of assembly scheme, the erratic holding fixtures, the incomplete operation, and so on. It is notable that for the assembly scheme, the locating method and the tolerance distribution program are mainly included.

The above inapposite settings should be corrected immediately. However, the over-adjustment might also cause the phenomena, and needs a theoretical guidance.

② When the measuring sample points fall outside the range of $[\mu-2\sigma,\mu+2\sigma]$ frequently, the phenomenon indicates that the standard deviation has an obvious large value. And it can be considered that the manufacturing and assembly process is abnormal. The common causes can be comprised as follows: the equipment/tools are not working properly in the assembly process, the locating and clamping accuracy of the end locators cannot meet the design requirements, KCs with a very loose engineering tolerance, or the distribution of required upper/lower tolerance limits should be relocated. With the purpose of solving the above problems, the positioning and clamping states of the tooling or equipment should be checked and recalibrated to meet the assembly requirements.

③ When the quality characteristic points distribute on both sides of the center line, the modes of step rise and step drop appear. This abnormity indicates that there is a change in the mean value of the statistical results, and there is a type of systematic error that causing the shifting. Although the steps in $X-R_s$ may indicate the assembly process was intentionally changed or improved, we analyze the distribution of quality data before beneficial solutions are taken. As a result, the conclusion is summarized that the systematic error is mainly contained by the changes of physical environment, such as temperature/gravity, the operator skill level, or the drilling/jointing deformation.

④ In the statistical control chart, when the distribution of the measurement points shows a noticeable continuous upward or downward trend, it can be considered that the manufacturing and assembly process is affected by continuous changes of the external factors, such as the wear of assembly tooling, loosening happened in assembly fixture, and inappropriate sampling frequency. The above factors would lead to the reduction of the quality in the manufacturing/assembly process. Process solutions that have the opposite effect should be taken, for amending the samples locations relative to the central line.

⑤ For one kind component of different batches, when plenty of the measurement points show a trend of increasing or decreasing at the same time, and demonstrate the same fluctuation status in the manufacturing/assembly process, then this situation indicates that the assembly quality is affected by a certain system error in one certain direction. The reasons may include factors such as operator, measurement system, assembly procedure, and locating/jointing devices.

⑥ During the assembly process among different batches, when the measurement values of the same characteristic/feature have a relatively fixed relationship with each other, it can be regarded that there is a deterministic system error occurred in the manufacturing process. Under this condition, the most important reason may be the inherent positioning deviation of assembly tooling/fixture, and it's necessary to re-calibrate the equipment for satisfying design requirements before starting the detailed assembly work.

By applying the above empirical criteria and solutions with a qualitative manner, i.e., the company's regulations on quality management rather than the theoretical calculation, it is easy to judge the abnormal reasons that exist in the detailed manufacturing/assembly processes of aircraft products. Then relevant adjusting strategies and improvement solutions can be put forward to optimize the assembly process.

6.6 Assembly Station Flowing Fluctuation Analysis at Different Time Stages for Multiple Components

With deviation control chart, by analyzing the quality characteristics during the manufacturing and assembly process, the SPC method can be used to distinguish whether the process is in a stable status. Then, with the detailed qualitative and quantitative analysis of the distribution and the trend of abnormal conditions, beneficial adjustment results can be gained for the specific assembly process. However, the above SPC charts mainly focus on the product quality characteristics under certain fixed assembly station, reflecting whether multiple quality samples are in a statistical stable state. Because of the small amount of the produced aircraft products, for the process deviation and fluctuation of a single quality sample under a statistical stable state, little attention has been paid on the quality data items at the same assembly station and the same assembly process, rather than at different time stages. In practical assembly, the values of specific type of quality sample at different assembly stations (i.e., the same assembly process at different time stages, such as parallel assembly lines) may also not the same and change dynamically over time. Then, the SPC chart would not reflect this situation. For the shape and dimension transferring process in aircraft manufacturing, in order to further analyze the fault source and improve the process flow, the concept of ASFF is proposed with the perspective of production statistics. ASFF is mainly fit for the deviation and fluctuation of assembly results under a certain assembly station at different time phases.

[Definition 1] assembly station flowing fluctuation (ASFF)

The core viewpoint of ASFF is similar to the PVTC that used in CNC machining process. In aircraft manufacturing process, with regard to one component at a certain assembly station across different time stages, ASFF refers to the expression of the trajectory diagram for the dynamic changes in the quality characteristics. And it is comprised of the fluctuation analysis of assembly quality samples. With ASFF, the quantitative evaluation for the capacity of flowing fluctuations during the whole assembly station can be achieved. As a result, the assurance ability of assembly station for product assembly quality is to be improved eventually.

To clarify the concept of ASFF, some extensional explanations of relevant terminologies are stated as follows.

(1) Deviation rate of assembly process

As is well known, the excessive deviation of the mean value during the assembly process would lead to product quality defects. Generally speaking, for the product assembly quality characteristics of the mean value and its design requirements, there exists a certain degree of deviation. A relative offset rate is proposed in order to reflect the deviation degree between them. And the deviation rate of assembly process can be denoted by Δd_{ij}, as shown in Equation (6-28).

$$\Delta d_{ij} = \frac{\overline{X}_{ij} - \mu_{ij}}{D_{ij}} = \frac{\frac{1}{N}\sum_{k=1}^{N} X_{ijk} - \mu_{ij}}{(USL_{ij} - LSL_{ij})/2} \tag{6-28}$$

In the above Equation, X_{ijk} represents the kth measurement value of the jth quality characteristic at the ith assembly station; $\overline{X}_{ij}$ represents their average value; μ_{ij} stands for the target value of the jth quality characteristic at the ith assembly station; USL_{ij} stands for the upper deviation limit; LSL_{ij} stands for the lower deviation limit; D_{ij} represents the centerline of the upper and lower deviation; and N stands for the number of measurement samples.

(2) Stability of assembly process

During the aircraft assembly process for a certain component, the fluctuation of quality sample has a significant influence on the consistency of different assembly errors and the stability of assembly quality. The Equation (6-29) can be used to describe the function effectiveness, with the relative stability coefficient of the manufacturing process, denoted as Δs_{ij}, to express fluctuation of the assembly station status flow.

$$\Delta s_{ij} = \frac{\sigma_{LT_{ij}} - \sigma_{ST_{ij}}}{\sigma_{LT_{ij}}} = \frac{\sqrt{\sum_{k=1}^{n}(X_{ijk} - \overline{X}_{ij})/(n-1) - R_s/d_2}}{\sqrt{\sum_{k=1}^{n}(X_{ijk} - \overline{X}_{ij})/(n-1)}} \tag{6-29}$$

With regard to the above equation, d_2 represents a constant value, d_2 is determined by the size of the sample group; and Δs_{ij} has a variation range value of [0,1]; $\sigma_{LT_{ij}}$

represents the long-term standard deviation; and $\sigma_{ST_{ij}}$ represents the short-term standard deviation.

[Definition 2] Trajectory Chart of Assembly Station Flowing Fluctuation

Considering the factors of the quality changes across different assembly time stages or different aircraft products, then for the deviation and fluctuation of the quality characteristics at a certain assembly station, form a fluctuation curve on the established fluctuation plane, namely, the station fluctuation trajectory diagram, as shown in Figure 6-7. The trajectory chart can be expressed by quality coordinates at different assembly stations. Where the quality coordinate is denoted by (Δd_{ij}, Δs_{ij}). Δd_{ij} stands for the value on the x-axis in the two-dimensional fluctuation plane, and Δs_{ij} is shown on the y-axis, respectively. Where the two fluctuation trajectories of two assembly stations at four different time phases are contained.

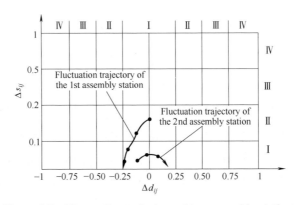

Figure 6-7　Fluctuation trajectory of two assembly stations

For the fluctuation trajectory of certain assembly station, it is notable that the division of fluctuation area for each station is determined by the six sigma quality theory. For the region Ⅰ in Figure 6-7, the horizontal axis and vertical axis stand for the normal offset of the deviation and a status that in a stable status, respectively. For the region Ⅱ, it represents the deviation is large and the quality status is less stable. However, for the regions Ⅲ and Ⅳ, the deviation becomes larger, and the quality status becomes more unstable. When some quality samples locate in regional Ⅱ or even in regional Ⅲ, it needs to solve the deviation problem at the current assembly station, i.e., it needs to further reduce the process fluctuation and deviation value. Other wisely, the assembly quality of the products would be poor and have an obvious inconsistency, and it may even cannot meet the design accuracy requirement.

The trajectory chart is firstly proposed in this section to reflect the quality

Digital Assembly Coordination and Quality Controlling Technology
for Aeronautical Thin-walled Structures

characteristics for a certain assembly station, i.e., the quality deviation and fluctuation state. With the analysis of the chart, the detailed improvment opportunities and strategies for each assembly stations can be acquired. As the quality data of the two assembly stations that shown in Figure 6-7, the deviation at the first station has a tendency to gradually increase, but the fluctuation degree has a tendency to gradually decrease. After analyzing these situations, it is found that the function effect of systematic errors in the assembly/manufacturing process is increasing, which means attentions should be paid to eliminate it. It is summarized that the systematic errors are mainly contained by the changes of physical environment, such as temperature/gravity, the operator skill level, or the drilling/jointing deformation. However, at the second assembly station, the deviation and fluctuating status are remain in the normal range, and the corresponding assembly operations are supposed to be maintained at the current level.

To summarize this chapter, for multi-components, by analyzing the fluctuation trajectory of a certain assembly station at different time stages, when the deviation value of the fluctuation state and the quality characteristic exceed the desired numerical values, some corresponding solutions should be taken. Beneficial results would be gained with the analysis on the quality trends that displayed intuitively, such as the assembly process at different time stages would be more stable, the fluctuation of quality characteristic is to be more and more consistent, and the error coordination consistency among different assemblies would be better. With the analysis of the large amount quality samples, the above optimization solutions would be more helpful in aircraft batch production.

6.7 Experimental Verification: the First Case

Assembly work of a typical wing-box component is taken as verification object to exhibit the effectiveness of the proposed method, i.e., reliability probability analysis and the adjustment strategies for assembly accuracy. Firstly, practical assembly scheme of the wing-box is described, and the data source for probability analysis is clarified. Secondly, the proposed methods and adjustment strategies are applied on this component. Finally, accurate optimization of repair quantity with actual conditions is carried out, and application effect is analyzed.

6.7.1 Assembly Scheme of the Wing-box Component

The aviation wing-box component often contains many complex structures, such as the integrated skin and segmented skin, the metal machining front walls and beams, hatch covers, and metal ribs. Many types of butt joints among skin panels and hatch cover components involve the heterogeneous laminated materials. For the traditional assembly technology, it's difficult to meet the quality requirements, such as strict gap clearance and step flush. In addition, the irregular curvature of the composite skin panels also makes it difficult to control the gap between skin panels and bearing frame parts. As a result, strategies based on active prevention, real-time control and prior repair compensation should be adopted to enhance the reliability of assembly deviation. The detailed assembly situation is shown in Figure 6-8, assembly fixture with accurate force and position adjustment functions is shown in Figure 6-9, and the relevant key data items are shown in Table 6-2.

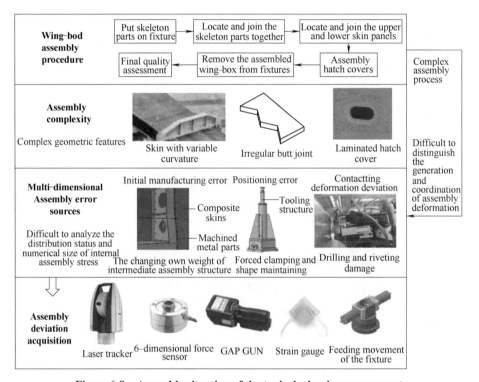

Figure 6-8   Assembly situation of the typical wing-box component

Combining with the detailed assembly fixture, to ensure the effectiveness of the reliability improvement method, other infrastructure conditions are also adopted. For the acquisition of key physical data during assembly process, measurement and sensing equipment, such as laser tracker, laser scanner, 6D force sensor, GAP GUN, and strain gauge, are contained. They could enhance the comprehensiveness and timeliness of assembly data collection. Then, during the transmission of assembly quality and control information, there are interactive information devices with the characteristics of high-speed and low delay, such as industrial field buses (CAN bus) and Ethercat field bus,

Digital Assembly Coordination and Quality Controlling Technology
for Aeronautical Thin-walled Structures

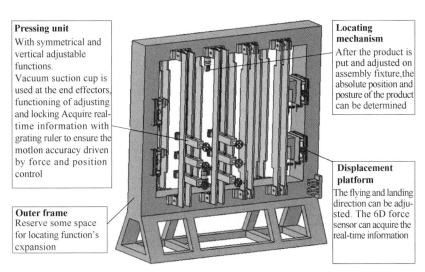

Pressing unit
With symmetrical and vertical adjustable functions.
Vacuum suction cup is used at the end effectors, functioning of adjusting and locking Acquire real-time information with grating ruler to ensure the motlon accuracy driven by force and position control

Outer frame
Reserve some space for locating function's cxpansion

Locating mechanism
After the product is put and adjusted on assembly fixture,the absolute position and posture of the product can be determined

Displacement platform
The flying and landing direction can be adjusted. The 6D force sensor can acquire the real-time information

Figure 6-9　Structure and description of the wing-box assembly fixture (See the color illustration)

Table 6-2　Key physical data items of assembly process scheme

Data items	Parameter type	Parameter value
Product	Contour profile	Length: about 2m; Width: About 1.3m; Height: About 0.3m
Assembly fixture	Motion stroke	X direction: ±70mm; Y direction: ±1000mm; Z direction: ±200mm
	Technical parameter	Resolution ratio of single motion direction: 0.05mm Positioning error: 0.05mm per total travel Repeat positioning error: 0.015mm per total travel
Measurement equipment	Laser tracker	Measuring range: about 40m; Accuracy: 0.03mm within 10m; Pitch angle: no more than 70°
	6D force sensor	Measurement range: 5000N
	GAP GUN	Accuracy: 0.001inch[①]; Maximum measurable clearance: 0.79inch; Maximum measurable step: 0.79inch

① 1inch=25.4mm.

to ensure the rapid transfer of the above assembly information. In addition, for the processing of assembly information, multi-core CPU processors with high-frequency, the storage media of large capacity hard disks with high-speed are adopted to ensure the efficient calculation and analysis of multi-source and heterogeneous spatial data. For the assembly scheme shown in Figure 6-8, the detailed processing methods can be described as: ① the initial manufacturing error is taken as the delivery profile variation of composite skin panels, machining frame, and other parts, ② fixture's positioning error is ensured by controlling the coaxiality of hinge joints and reference holes with vacuum suction cup (3-DOF), ③ joining quality is ensured by automatic drilling and riveting equipment with the optimized parameters, ④ force and position information is obtained by 6D force sensor mounted on fixture, laser tracker, and others, ⑤ internal stress state is measured using strain gauge that mounted on skin panels and fixture, ⑥ feed movement value of fixture's adjustment and the drilling and riveting operations are obtained by interactive devices that previous mentioned. Then the deviation coupling value would be obtained according to the stress action of various geometric error sources.

6.7.2 Active Prevention and Real-time Adjustment Control on Assembly Out-of-tolerance Event

For the detailed assembly procedures of wing-box component, the verification process is summarized as follows. Before the practical assembly work, with the measured error items of parts, assembly fixtures, and actual assembly conditions at previous process links, analysis work is carried out to predict assembly accuracy and repair amounts accurately. Important preventive optimization measures are taken, such as calculating the out-of-tolerance probability for skin assembly profiles, gaps and steps during the entire assembly process, and simulating the repair compensation efforts before actual assembly operations are carried out on the ending ribs. It is mentioned that in virtual simulation environment, only the analysis results could fulfill required quality indexes, allowing the practical assembly work to be proceeded subsequently. On practical assembly site, because the detailed geometric and physical variables are measured or sensed with different kinds of measurement and sensing equipment, as well as information processing mediums. With preventive actions, the prior analysis results could be considered to have high accuracy. To be more specific, firstly, error sources of internal assembly stress and their impact on product's final geometric deviation are analyzed. Then, with the accuracy probability calculation that driven by dynamic sliding window method, they are realized by the out-of-tolerance prevention and optimization system (Figure 6-10). The system takes online calculation of EOCDCF probability as the core, and the imported assembly data and deviation coupling function are taken as analytical data source. FEA analysis is also adopted as the auxiliary confirmation of calculation results.

Considering the actual measured error data and practical assembly conditions, the detailed verification process in Figure 6-10 can be described as follows.

① The selection function of dynamic sliding window is the premise of acquired data's calculation. It's realized mainly by identifying key assembly procedures (such as locating and clamping ending wing ribs, drilling and riveting operations for wing ribs and composite skins) and key geometric features (such as skin profiles with variable curvature, position of ending wing rib's web plate). In addition, the increasing measurement density and improving frequency of data acquisition, would also be more effective.

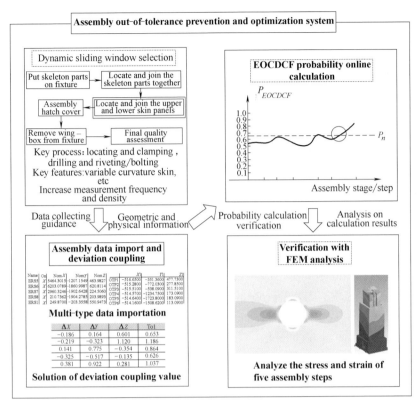

Figure 6-10　Schematic diagram of assembly out-of-tolerance prevention and optimization system

② For the numerical solution of deviation coupling, it's mainly based on collecting and summarizing the assembly data, which is timely imported into the system with communication methods such as field bus. The collected multi-type assembly process data includes key measurement points such as ERS (enhanced reference system) points arranged on fixture's frame, reference TB (tooling ball) points arranged on the ground, and POTP (product optical tooling point) designed on the parts to be assembled.

③ As the core of this system, the online calculation function of EOCDCF probability mainly combines the assembly deviation data to realize the numerical calculation of its occurrence probability P_{EOCDCF}. The assembly steps of upper and lower skin panels are selected as an example. Where, the probability threshold P_n for the current process is set as 0.65 according to engineering experience. Then taking the assembly process with dynamic sliding window as an analytical case, the probability calculation process is briefly described. According to the measurement density at this assembly stage, assuming the fixed sliding window size N_d is 90, the dynamic sliding window size N_j is 10, the prediction gap size is 100, and the forgetting factor is 0.8. After the deviation coupling data is introduced, the error's variation expectation g_j which is preliminarily calculated, is smaller than the acceptable average deviation growth b_j, indicating no obvious out-of-tolerance phenomenon. Then, with the further analysis of the coupling data, the obtained number N_l that have out-of-tolerance cases in the fixed sliding window is 30, while the corresponding number N_{lj} in the dynamic sliding window is 4. According to the above parameters, the preliminary estimation $\eta_f(n)$ can be gained according to Equation (6-15). After the probability estimation $\hat{\pi}_h(n)$ is calculated with the forgetting factor and Equation (6-13), the calculation result can then be obtained according to the moment generating function $M(\lambda)$, the rate function $I(b_j)$ and Equation (6-11).

After the above calculation operations, the probability calculation results at this assembly stage can be represented as a curve considering probability and assembly time, i.e., the curve under "EOCDCF probability online calculation" in Figure 6-8. Where, the situation of $P_{EOCDCF} \geqslant P_n$ appears, and after the assembly procedure reaches the above Step 5, i.e., removing the assembled wing-box from assembly fixture, P_{EOCDCF} behaves a continuous upward tendency, exceeding 0.8 at one point. With preliminarily analysis, the reason is the springback deformation occurring as assembling variable curvature skins, which makes the actual assembly deviation difficult to meet design expectations. The occurrence of EOCDCF event should be further confirmed with FEA simulation.

④ For the FEA verification function, the stress-strain analysis is mainly carried out on assembly steps that have an occurrence of $P_{EOCDCF} \geqslant P_n$. In Figure 6-8, the FEA simulation is performed, especially for the obviously rising area of P_{EOCDCF} in the joining process. However, the results show that the stress and strain of related fixtures and parts are still at a high level, and there are potential factors for the occurrence of EOCDCF event.

⑤ Finally, according to the results obtained by this optimization system, assembly procedures and work steps that are prone to lead to the EOCDCF event would be fed back to the technicians and workers, and then, corresponding improvement measures would be taken. In detailed, before the assembled component is removed from assembly fixture, even the FEA verification method has been adopted, the out-of-tolerance phenomenon, that P_{EOCDCF} exceeds 0.8 in Step 5, still happened. Then, the accuracy's reliability adjustment strategy of Mode 3 in Table 6-1 is taken in advance according to the probability threshold relationship. To be more specific, the adjustment strategies can be shown in the following four aspects.

Digital Assembly Coordination and Quality Controlling Technology
for Aeronautical Thin-walled Structures

a. As for the adjustment of the assembly process parameters, according to the guidance of the process parameter library, measures such as reducing the joining feedrate from 300mm/min to 100mm/min are adopted, to enhance the robustness of the assembly process.

b. In terms of checking the product's position and shape before it's removed from assembly fixture, the accuracy data of key parts is analyzed by collecting the data from key product features, fixture's measurement points, and datum reference points. After the comparisons and analysis of theoretical models, it is concluded that there is an excessive deformation in skin panels. And then for the corresponding pressing unit and locating mechanism of the tooling system in Figure 6-7, adjustment to the force and position should be adopted to realize the checking and guaranteeing of part's positions and shapes.

c. For assembly information acquirement and fixture's adjustment during the removing process from assembly fixture, the measurement frequency of laser tracker, 6D force sensor, and other sensing equipment should be improved, and the measurement density of key product points distributed on variable curvature skin panels, butt joints, and fixture's key structures (such as tooling pressing unit) should be increased, and the data transmission efficiency should also be improved.

d. For the loading and unloading sequence of clamping force, factors such as material type, stiffness, and presence of opening holes, variable cross-sections, and curvature features in the structures are taken as the judgment principles for the sequence design. To the loading/unloading of clamping force for parts and components of composite materials and parts with poor rigidity and complex structures are given priority.

6.7.3 Accurate Optimization of Repair Compensation Based on Measured Data and Assembly Conditions

For the above wing-box component, the assembly accuracy of skin profile, assembly gaps, steps, and other key design requirements are often difficult to guarantee, and the assembly out-of-tolerance phenomenon is easy to occur. The low success rate of one-time assembly would cause a large amount of manual repair works and low assembly efficiency during actual production. To improve its accuracy reliability, assembly repair solutions have to be taken for keeping the final geometric accuracy and internal stress within the controllable range. Then, in order to gain the detailed repair areas and repair quantities before the final assemblies process, strict requirements are required for the assembly procedures, and position deviation of the ending ribs is specified within a range of [−0.5, 0.5]mm. In this section, the control on position deviation of ending ribs are taken as an example, in which the repair scheme is generated in advance. To be more specific, this section applies the method for improving the reliability of assembly accuracy from the following three steps.

① Reconstruct part's model. The mating and joining surfaces of the two ending ribs, mating surfaces of the front/rear beams, and the upper/lower skin panels' assembly are selected as the key features. With laser tracker, a portion of coordinate data is shown in Figure 6-11. Geometric reconstruction is carried out with the measured data, and the deviations between the point clouds and the reconstructed surface are detected by the "Deviation Analysis" function. As shown in Figure 6-12, the reconstructed surface of the mating area for the front beam, and the joining area of upper skin panel and ending ribs are shown. The maximum reconstruction deviation is about 0.06mm.

P118,	20738.077,	-932.078,	793.810	P50,	18302.651,	-771.473,	-1947.583
P119,	20729.724,	-1030.582,	777.159	P51,	18202.871,	-776.533,	-1953.632
P120,	20721.305,	-1129.522,	763.712	P52,	18055.330,	-789.326,	-2049.155
P121,	20712.837,	-1229.063,	755.109	P53,	18155.034,	-780.751,	-2054.263
P122,	20704.347,	-1328.707,	753.336	P54,	18254.752,	-773.975,	-2049.583
P123,	20695.822,	-1428.496,	753.259	P55,	18354.509,	-768.767,	-2054.723
P124,	20687.287,	-1528.295,	756.492	P56,	18454.734,	-764.821,	-2053.519
P125,	20678.812,	-1626.912,	771.976	P57,	18554.796,	-762.196,	-2050.391
P126,	20672.360,	-1701.305,	838.550	P58,	18654.774,	-760.599,	-2051.932
P127,	20673.204,	-1690.508,	938.101	P59,	18753.878,	-759.494,	-2065.376
P128,	20678.626,	-1626.591,	1015.077	P60,	18853.515,	-759.233,	-2075.482
P129,	20687.037,	-1528.315,	1032.388	P61,	18953.602,	-760.201,	-2074.941
P130,	20695.562,	-1428.324,	1030.041	P62,	19053.675,	-762.008,	-2073.585
P131,	20704.055,	-1328.385,	1031.750	P63,	19153.457,	-763.975,	-2080.512
P132,	20712.582,	-1228.625,	1035.125	P64,	19253.171,	-766.749,	-2087.678
P133,	20721.115,	-1128.757,	1041.089	P65,	19352.175,	-769.499,	-2102.953
P134,	20729.475,	-1031.044,	1062.152	P66,	19451.403,	-773.008,	-2116.784
P135,	20737.179,	-941.105,	1105.337	P67,	19551.022,	-778.831,	-2109.541
P136,	20737.250,	-939.503,	1205.409	P68,	19650.893,	-785.342,	-2103.413
P137,	20738.273,	-927.276,	1304.714	P69,	19749.526,	-793.347,	-2088.535
P138,	20729.990,	-1023.022,	1333.038	P70,	19849.089,	-801.636,	-2080.643
P139,	20721.379,	-1122.808,	1325.944	P71,	19948.855,	-809.547,	-2084.301
P140,	20712.921,	-1221.380,	1308.514	P72,	20048.693,	-818.023,	-2087.565

Figure 6-11   Measured data of the key features

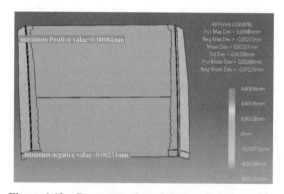

Figure 6-12   Reconstruction of the key feature surface

② Predict ending rib's position deviation. The reconstructed model is imported into 3DCS software, and deformation of upper skin panel caused by its own gravity is calculated. The specific procedures are: a. import geometric models of skin panel, locating fixture, and clamping device into 3DCS software to generate assembly process model; b. import simplified assembly model into ABAQUS software, to define material properties and assembly conditions; c. import mesh file into the FEA module of 3DCS,

Digital Assembly Coordination and Quality Controlling Technology for Aeronautical Thin-walled Structures

then generate stiffness matrix and calculate the panel's deformation. With above operations, assembly constraints of face-to-face mating for ending ribs are then defined on actual surface. Finally, the assembled position deviation of each measuring points is calculated with Monte Carlo simulation, and their results are represented by the box diagrams in Figure 6-13 and Figure 6-14. It's known that the position deviation of ending ribs does not meet design requirements, and its quality needs to be further ensured with repair compensation actions.

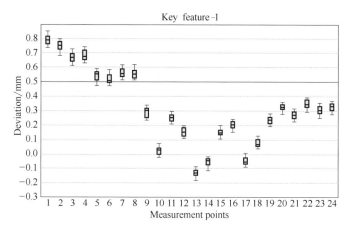

Figure 6-13 Assembly position deviation of left rib on different measuring points

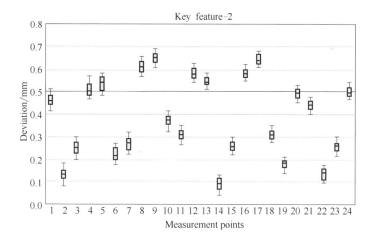

Figure 6-14 Assembly position deviation of right rib on different measuring points

③ Generate repair scheme. According to the repair dimension ring selecting rules, for the front and rear beams, the two ends of upper surfaces are chosen as repair areas, as shown in Figure 6-15. The above four areas have a simple shape and are not common error links. Position deviation of left ending rib is guaranteed by repairing areas No. 1 and No. 2, while for the right ending rib, that are ensured by areas No.3 and No.4. By calculating assembly deviation of the above surface, the repair amount of each repair area is determined by IPSO optimization method to generate the most feasible repair scheme.

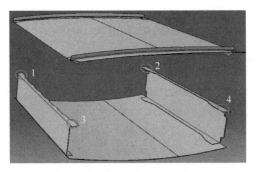

Figure 6-15   Schematic diagram of repair areas

For the repair scheme to control the assembly position deviation of the left and right ending ribs, the traditional repair documents state only the description of "the allowable shimming amount is less than 1mm and the grinding amount is less than 0.5mm". It does not involve the specific repair area and repair amount values, and also cannot guide the repair area's distribution overall, which would make the workers rely on their personal experience and repeated manual operations. Correspondingly, the assembly efficiency is low. However, for the solution in this section, by analyzing the geometric mating status of multiple repair areas, the specific repair values required for different assembly areas are clearly given in the generated repair scheme file, which could avoid the repairing blindness. In addition, by determining several groups of alternative repair schemes and ranking them according to evaluation methods, such as fuzzy comprehensive solution, it can be more applicable for the practical engineering situation. In summary, through the modeling calculation, the optimal repair scheme obtained in this section is: area No.1 is grinded by 0.2mm, with the shimming amounts in areas No.2 and No.3 of 0.6mm and 0.4mm, respectively, and no repair is required in area No.4. It can be seen the repair amount of each area is within the allowable repair range.

6.7.4 Application Effect of Assembly Accuracy Reliability Improvement Method

For the final assembly quality index evaluation, the data obtained from practical assembly site are compared with the final measurement data without combining preventive optimization and precise repair compensation solutions, as shown in Table 6-3. Where, the measurement device for obtaining the assembly gap and step data in this table is the gap gun that shown in Figure 6-5. The statistical analysis results show that the gap and step quality between the skin butt joints are increased by 28.60% and 28.86%, respectively. Through the repair simulation and accurate compensation operations, the deviation of joining surface for ending ribs is increased by 41.42%, and there is no extreme assembly out-of-tolerance phenomenon by adopting the above benefit adjustment actions. From the micro perspective, the above engineering application efforts indicate that the research methods could generate the accurate repair schemes in advance, and the target assembly accuracy could also meet the assembly requirements. In practical assembly site, the blindness of repair operations could be eliminated, the cost and difficulty of repair work could also be reduced, and the assembly efficiency is also increased. From the macro perspective, considering the cumulative accumulation process of different assembly error sources, it is particularly important for the assembly work of other complex, thin-walled products, and the solution also has an obvious beneficial affection on product's service performance.

Table 6-3   Comparison of assembly data before and after using the accuracy reliability improvement measures   unit: mm

Assembly quality index	Before optimization	After optimization	Quality improvement statistics
Assembly gap between skin profiles≤0.5	Distribution range: [0.452, 0.659] (Statistical mean value: 0.584)	Distribution range: [0.357, 0.495] (Statistical mean value: 0.417)	(0.584−0.417)/0.584× 100%=28.60%
Assembly step between skin profiles≤0.3	Distribution range: [0.285, 0.486] (Statistical mean value: 0.395)	Distribution range: [0.254, 0.298] (Statistical mean value: 0.281)	(0.395−0.281)/0.395× 100%=28.86%
Outer profile of ending ribs≤0.5	Distribution range: [−0.155, 0.783] (Statistical mean value: 0.536)	Distribution range: [−0.394, 0.0724] (Statistical mean value: −0.314)	(0.536−0.314)/0.536× 100%=41.42%
Remarks	Draw the out-of-tolerance sheet and discuss whether to scrap the current component	Meet the required accuracy requirements	The reliability of assembly accuracy is significantly improved

6.8   Experimental Verification: the Second Case

To verify the feasibility in Sections 6.5 and 6.6, the structure and assembly process of the experimental object can be shown in Section 2.4.

6.8.1   Quality Data Measurement Sampling of PCFs

According to the design requirements and the official assembly document, the flush coordination accuracy between skins of inner component and outer component should be guaranteed in the assembly process. The flush is defined as the offset or altitude difference between the two different skin profiles. Considering that the 1st rib and the 16th rib have the similar: ① geometric dimensions, ② geometric shapes/profiles, ③ the required assembly profile tolerance, and ④ the specific positioning or assembly methods, then take another five factors in account: ① the left and right components are exactly symmetrical relative to the design datum of the aircraft, ② the installation and adjustment works of the assembly jig is completed by the same band of workers, ③ the assembly process level for the technicians that responsible for the inner and outer components is roughly the same, ④ the assembly work starts at the same time, and ⑤ the assembly tooling and assembly environment is approximately the same for different components. Therefore, the assembly analysis work can be done. It is noteworthy that the above nine assumptions correspond to the regular on-site assembly operations. By integrating these nine factors comprehensively, a conclusion can be drawn, namely, the manufacturing normative requirements of the four components are basically the same. As a result, the measurement and test data for the skin shapes of the ending ribs of the four components, can be combined together to construct the simulated sample data in the statistical control process. Correspondingly, the number of samples can also be expanded, which also means there are same kind of quality data at different assembly time stages or stations. In the practical production, by recording the measured skin profile data of the four ending ribs of five different aircraft products, i.e., the 01st, 03rd, 04th, 06th and 08th, a simulated sample that containing twenty data points is constructed. And with the Equations presented in the Section 6.2, the samples' statistical data information are calculated and analyzed, as shown in Table 6-4.

Table 6-4   Statistical data information of the profile deviation for the ending ribs

unit: mm

Quality sample	01st flight				03rd flight				04th flight	
	1	2	3	4	5	6	7	8	9	10
X	0.36	0.38	0.40	0.08	0.27	0.35	0.32	0.33	0.35	0.39
R_s		0.02	0.02	0.32	0.19	0.09	0.03	0.01	0.02	0.04
Quality sample	04th flight		06th flight				08th flight			
	11	12	13	14	15	16	17	18	19	20
X	0.29	0.31	0.18	0.24	0.29	0.17	0.28	0.26	0.35	0.20
R_s	0.10	0.02	0.13	0.06	0.05	0.12	0.11	0.02	0.09	0.15

In practical assembly site, it is notable that the measuring equipment for collecting the sample data of the skin profile is a laser tracker, namely the API III. To measure the above source data and define the profile deviation with the data from laser tracker, seven important steps are adopted and described in detail, as shown in Figure 6-16.

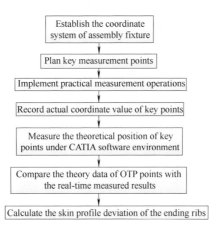

Figure 6-16 **Measurement procedure on the profile deviation for the ending ribs**

Step 1 Establish the coordinate system of assembly fixture. In the assembly system, there are four TB datum measurement points located at the corners of the adjustable support foundation. The first three TB points are adjusted to the same height with a special device, called gradienter. Then, the CATIA model of the assembly fixture is imported to measurement software SA (spatial analysis) that is exclusively used with laser tracker. And this software can streamline the process of evaluating the deviations and their minimizations, such as best fit. After fixing the position of TB1, with the method of shaft alignment, the position of TB2 is adjusted along y-direction. Correspondingly, the X-axis of the fixture coordinate system is determined. With the same method, after the position of TB3 along x-direction is fixed, the Y-axis is also determined. For the Z-axis, it can be determined automatically with the guidance of right-hand rule, based on the obtained X/Y-axes. Thus, the Cartesian coordinate system of tooling system, i.e., $OXYZ$, can be automatically generated in the laser tracking measurement system. It is noteworthy that the fourth TB point has a function of checking and adjusting the built reference system. Then, the assembly measurement work can be carried out.

Step 2 Plan key measurement points. For the skin profiles of the ending ribs, due to the surface curvature does not change drastically, 22 key OTP measurement points are planned in practical assembly site, as shown in Figure 6-17. The method of equidistant measurement is adopted, with the distance between two adjacent points of 1/11 of the part's length. These points are distributed uniformly along the flight direction in two rows, nearly covering the whole regions of skin profile at the ending rib.

Step 3 Implement practical measurement operations. In the practical assembly environment and in the tooling coordinate system, for each ending rib, measure the position of each key OTP point [$P_i = (x_i, y_i, z_i)$, i=1, 2,$\cdots$, 22] ten times with a laser tracker.

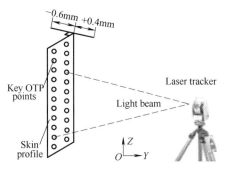

Figure 6-17 Key measurement points planning at the ending rib

Step 4   Record the actual coordinate value of key points. The average coordinate value of each P_i is noted as $\bar{P}_i = (\bar{x}_i, \bar{y}_i, \bar{z}_i)$, which is taken as the actual coordinate position status, with the purpose of minimizing the measurement error.

Step 5   Measure the theoretical position of key points under CATIA software environment. The mathematical model of the components is imported into the CATIA virtual assembly environment, where the theory coordinates of key measurement points that distributed on assembly platform can be gained in the virtual software. Where, the theory coordinate value of each OTP point is recorded as $P_{i0} = (x_{i0}, y_{i0}, z_{i0})$.

Step 6   Compare the theory data of OTP points with the real-time measured results. Calculate the difference between $\bar{P}_i$ and P_{i0}, recording the result as $\Delta P_i = (x_i, y_i, z_i)$.

Step 7   Calculate the skin profile deviation of the ending ribs. The profile deviation is taken as the average $\Delta P_i = (x_i, y_i, z_i)$ of all the key OTP points, i.e., $\Delta P = (\sum_{i=1}^{22} \Delta P_i) /$ $22 = (\Delta x, \Delta y, \Delta z)$.

For the measurement results in Table 6-1, there are three aspects need to be explained.

● It is known that the profile deviations along the vertical direction has the most important influence on the flight performance. As a result, only one value, rather than three, is taken as the representation of the entire assembly deviation of the skin profile.

● The positive measurement value indicates the skin profile is larger than the theoretical shape, while negative value indicates it is smaller.

By browsing the assembly technique documents, the desired assembly deviation has a range of [−0.6mm, 0.4mm]. Combining with the measurement characteristics of laser tracker, the deviation results in Table 6-1 are displayed as two decimal places.

6.8.2   Assembly Quality Control with SPC and Feedback Improvement Actions

Based on the above statistical data listed in Table 6-4, the individual value range

Digital Assembly Coordination and Quality Controlling Technology
for Aeronautical Thin-walled Structures

control chart, i.e., $X - R_S$, is drawn according to Equation (6-1)~(6-4), as shown in Figure 6-12 to Figure 6-14, Section 6.2 and Section 6.7.

As shown in the individual control chart, for the ending rib of the left outer wing flap in the first measurement data set, it can be observed that the position of the skin profile measuring points fall within the upper and lower control lines. Then by observing the moving range control chart, it can be seen that the measuring results of the profile deviation for the ending ribs exceed the upper and lower control lines. Based on the distribution form of the above quality samples, it can be judged that there are abnormal situations existing in the manufacturing and assembly process. The improvement solution can benefit from the first criterion, "Analysis on SPC chart in aircraft assembly with qualitative practical experience". And the major shortcomings of the assembly scheme are analyzed.

As a matter of fact, in the practical assembly work of the wing components, the ending ribs are located with the help of a dedicated rigid assembly tooling. More specifically, the ribs are positioned by the positioning block that mounted on the counter boards along the spanwise direction, and the ribs are also positioned by a movable block, with a gap of skin thickness along the flying direction. It is noteworthy that the position of the movable block is adjusted manually by taking the actual precision or position status of the ribs into account. The above operation is carried out according to the worker's experience. This manual method results in the limitations of design skills on the rigid assembly jigs from about 20 years ago. And the traditional analog assembly technology has a low digitization degree. As assembling the wing flap component, shown in Figure 6-18, positioning method based on contour boards is often adopted. Where, f_1 represents the rib's profile; f_2 represents the contour board; f_3 represents the theoretical contact surface between rib and contour board; f_4 represents the actual contact surface between the above two kind of parts, and ∇ represents the assembly error between the ending rib and its positioning tooling. Although the assembly equipment has achieved the predefined

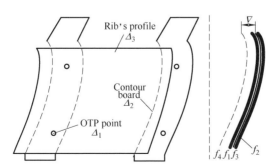

Figure 6-18   Positioning method based on contour boards

accuracy and stability before it has been used in the assembly process, the behaviors caused by the assembly operations, such as assembly deviation and deformation resilience, make it always difficult to guarantee the assembly quality of complex thin-walled structures.

In order to address the above problem and decrease the fluctuation range, by taking the detailed solution, an efficient action is taken in practical assembly site: a special positioning device for each end wing rib is designed and added, as shown in Figure 6-18. Considering the mating holes has a simpler assembly relationship than the profiles of contour boards, the positioning method based on coordination holes is proposed. Coordination holes that distributed on the rib and the locator have the same dimensions and positions in the unified coordinate assembly system, and they appear in pairs. Front area of the two flanks of the locator is used for locating the rib's web, which includes two coordination holes. The coordination holes can also be used as optical target points for locating the rib with a laser tracker. The above assembly method, i.e., one-surface-two-holes, can improve the assembly accuracy and avoid many of the assembly problems with contour boards. The assembly statistical result of the following four data sets showe that the assembly quality is obviously improved by taking the above solution.

For the individual control chart, i.e., Figure 6-17, it can be known that, for the assembly work of the first three products, the position of most measuring points distributed on the skin profile falls above the center control line. Considering the lower assembly deviation control limit, i.e. −0.6mm, its absolute value is greater than the upper assembly deviation control limit, i.e., 0.4mm, solutions should be taken in order to shift the positions of measuring points in the chart toward the lower limit, with a lower exceeding risk. And an efficient solution is taken to decrease the deterministic systematic errors, with regard to the above phenomenon. To be more specific, the solutions can benefit from the criterion that described in Section 6.5. In practical assembly site, for the mating process between the rib and the spar that shown in Figure 6-13, there is a phenomenon that the rib often has a position deviation in the flying direction. The above position deviation would cause a large coaxiality error between ① the mounting holes on both ends of the counter boards, and ② the locating hole on the bracket of the assembly jig. Then, the locating pins that passing through the above two holes would be in an inflexible rotational state. In extreme cases, the pins couldn't be inserted or become block. To solve this problem, the quantitative optimization solutions that described in Figure 6-3 are taken into account, with continuous improvement actions, such as ① modeling the coordination relationship of pin-to-hole according to the practical error status, ② calculating the error chain of skin profile by taking all of the error items into account: the manufacturing error, the tooling error, the assembly error, etc., and

Digital Assembly Coordination and Quality Controlling Technology
for Aeronautical Thin-walled Structures

③ diagnosing the coordination error sources with a hierarchical, cooperative way.

Then, the tolerance information between the rib and the spar can be redesigned and controlled within an acceptable value with an active feedback method, as shown in Figure 6-19. By browsing the official assembly documents, the machining error of the two mating parts is ±0.15mm. According to the CRs of the corresponding geometric features, the assembly precision of key OTP points on the rib profile can be calculated by Equation (6-30). Where, α stands for rotation angular deflection at the sleeve mating area (denoted by the dotted line).

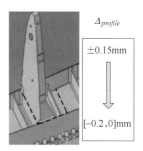

Figure 6-19 Optimization of the tolerance information between the rib and the spar

$$\Delta_{profile} = \Delta_{rib-spar}\Delta_\alpha = [-0.029, 0.171]\text{mm} \qquad (6\text{-}30)$$

It can be known the calculated result takes a large portion of the desired assembly deviation range. To loose the tolerance range of other error links, with the instruction of Sections 6.5, the error of the above two parts is optimized to [−0.2, 0]mm. With this solution, the locating pins have a flexible rotational state, and the quality data of the 4th and 5th products show a massive opposite tendency, which means the assembly quality is improved.

To sum up, with the above optimization actions, firstly, the locating pins that working with the counter boards would have a relatively flexible rotation performance. And then, for the assembly work of the following aircraft products, the assembly error becomes easier to guarantee. With another aspect, the measuring points of the skin profile at the position of the ending ribs, i.e., the 4th and 5th data samples, would fall below the central control line, and are also closer to the centerline of required assembly deviation range. In practical assembly site, similar solutions are also adopted to the mating area of other ribs and spars, and the skin profile deviation can fit the design requirements of both the theory analysis and practical measurement.

6.8.3 Assembly Quality Control with ASFF Analysis for Multiple Components

For the purpose of guaranteeing the assembly quality, other solutions are also taken using ASFF analysis. For the assembly work of certain products, relevant analysis proceedings on the skin profiles of the ending ribs for the four components have been done. The dynamic changes of the assembly quality at different time stages is taken into account. Before adding the specific locating device for the ending wing ribs and optimizing the tolerance information between the rib and the spar, the quality characteristics data for each coordination feature at six different time periods are collected, i.e. the deviation of the skin profile samples at the ending ribs, as shown in Table 6-5.

Table 6-5 Statistical data of skin profile at the ending ribs
(Without assembly quality control actions)   unit: mm

Quality sample	1	2	3	4	5	6
Q_{in-r}	0.28	0.18	−0.21	−0.35	0.24	0.34
Q_{out-r}	0.26	0.25	−0.19	0.18	−0.27	0.17
Q_{in-l}	0.35	−0.24	0.13	−0.26	0.18	0.24
Q_{out-l}	0.20	−0.22	0.24	0.14	0.23	0.18

As illustrated earlier in Table 6-5, where Q_{in-r} stands for the quality characteristic of the right inner component, Q_{out-r} represents the quality characteristic of the right outer component, Q_{in-l} represents the quality characteristic of the left inner component, Q_{out-l} stands for the quality characteristic of the left outer component. Through the calculation works shown in Equation (6-18) and (6-19), the distribution results of the above four quality characteristics' fluctuation situation can be shown by the dark-colored dots in Figure 6-20, located in the upper region of this graph.

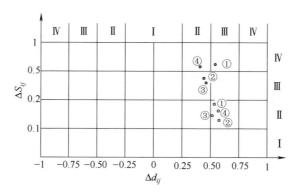

Figure 6-20 Assembly Station Flowing Fluctuation diagram of the wing components

Through observation on Figure 6-20, according to the analysis in the Section 6.6 "Assembly station flowing fluctuation analysis at different time stages", it can be known that the process deviation of the above four quality features is large, and the assembly process is unstable. However, after adding the special positioning locators that have a locating function for the ending wing rib, and optimizing the assembly parameters, the practical data of each quality features that collected at six different assembly periods is shown in Table 6-6. After adding the above actual data into Equation (6-28) and (6-29)

Digital Assembly Coordination and Quality Controlling Technology
for Aeronautical Thin-walled Structures

for calculation, the fluctuation results can be shown by the light-colored dot in Figure 6-18, located in the lower region of the graph.

Table 6-6   Statistical data of skin profile at the ending ribs

(With assembly quality controlling actions)   unit: mm

Quality sample	1	2	3	4	5	6
Q_{in-r}	−0.35	−0.22	−0.21	−0.19	−0.34	−0.25
Q_{out-r}	0.17	0.35	0.28	0.34	0.33	0.19
Q_{in-l}	−0.25	−0.34	−0.21	−0.19	−0.31	−0.22
Q_{out-l}	−0.22	−0.24	−0.18	−0.29	−0.36	−0.40

With the analysis work that mentioned above, it can be concluded that after taking the assembly quality control actions, the average process deviation of quality features still have a large numerical value, ranging from 0.128mm to −0.134mm. And further optimization and improvement actions should be done, such as the redesign and relocation of the part design errors, the tooling installation/locating errors, the assembly deformation errors, datum transformation errors. However, to show the optimization results, the deviation rate of assembly process Δd_{ij} changed from 0.473 to 0.525, with a fluctuation range of (0.525−0.473)/0.473×100%=10.994%, which is acceptable in practical assembly site. The stability of assembly process Δs_{ij} changed from 0.46 to 0.16, with an increase range of (0.460−0.160)/0.460×100%=65.217%, which means the optimized assembly process is more stable, and the assembly quality is improved by adopting the above beneficial actions.

6.9   Summary

With the research and application of reliability improvement methods for assembly accuracy with out-of-tolerance probability analysis, prior precise repair optimization, SPC, and ASFF analysis, the following conclusions are gained.

① For the non-independent relationships among multi-dimensional error sources, their internal stress function with practical error items has a strong influence on the EOCDCF phenomenon.

② The probability method based on deviation analysis and LDP theory is established, which is effective for evaluating and improving accuracy reliability, with feedback adjustment on reliability results that could decrease the occurrences of EOCDCF event.

③ Considering the actual deviations and assembly conditions, once the EOCDCF phenomenon has occurred, the detailed repair areas and quantities for parts would be designed accurately in advance with the proposed two optimization process.

④ With SPC analysis, assembly feedback actions, and ASFF analysis, the assembly quality control in practical production process is completed, with some specific beneficial improvement actions put forward through the comprehensive use of the above qualitative and quantitative methods.

⑤ Engineering application results show that the proposed method can enhance the ability of assembly process accuracy in the controlled range. Through precisely controlling the assembly process and generating accurate repair plans in advance, the target assembly quality improvement rate is about 30%, and the assembly reliability shows an significant enhancement.

⑥ Although the solution in this chapter could improve assembly accuracy and internal stress, enhancing product's service performance. However, the following three aspects should be further concerned.

a. In terms of efficiency and accuracy of online probability calculation, it is necessary to determine the selection of either a dynamic or fixed sliding window, considering the actual measurement and sensing frequency of key assembly processes. From another aspect, for the efficiency of online adjustment, in order to decrease excessive process adjustment, a reasonable probability threshold should be further determined according to the contribution of each process links on assembly quality.

b. In evaluating the repair scheme, only the influence of repair scheme on target assembly accuracy is considered, however, the influence of shimming and grinding operations on the strength, fatigue and other performance indexes of assembled structures are not considered.

c. Considering the benefit effort of digital twin technology, it is necessary to further combine the idea of "virtual-reality integration, control reality with virtual simulation" to establishment a closed-loop controlling mechanism on improving assembly quality.

References

[1] Lupuleac S, Pogarskaia T, Churilova M, et al. Optimization of fastener pattern in airframe assembly[J]. Assembly Automation, 2020, 40(5): 723-733.

[2] Haeger A, Grudenik M, Hoffmann M, et al. Effect of drilling-induced damage on the open hole flexural fatigue of carbon/epoxy composites[J]. Composite Structures, 2019, 215(5): 238-248.

[3] Taghizadeh H, Chakherlou T. Fatigue behavior of interference fitted Al-alloy 7075-T651 specimens subjected to bolt tightening[J]. Proceedings of the Institution of Mechanical Engineers, Part L: Journal of Materials: Design and Applications, 2019, 233(9):1879-1893.

[4] Bullen G. Automated/mechanized drilling and countersinking of airframes[M]. Warrendale: SAE International Press, 2013.

[5] Yin L, Tang D, Ullah I, et al. Analyzing engineering change of aircraft assembly tooling considering both duration and resource consumption[J]. Advanced Engineering Informatics, 2017, 33: 44-59.

[6] Zhu D, Zhang Z, Shi L, et al. A hierarchical assembly knowledge representation framework and microdevice assembly ontology[J]. Advanced Engineering Informatics, 2022, 53: 101705.

[7] Huo J, Lee G. Intelligent workload balance control of the assembly process considering condition-based maintenance[J]. Advanced Engineering Informatics, 2021, 49: 101341.

[8] Li S, Tang D, Xue D, et al. Assembly sequence planning based on structure cells in open design[J]. Advanced Engineering Informatics, 2022, 53: 101685.

[9] Polini W, Corrado A. Digital twin of composite assembly manufacturing process[J]. International Journal of Production Research, 2020, 58(17): 5238-5252..

[10] Lockheed Matin. How to Build Aircraft Articles in Half the Time[EB/OL]. (2021-5-12) [2024-12-03]. https://www.lockheedmartin.com/en-us/news/features/2021/How-to-Build-Aircraft-Articles-in-Half-the-Time.html

[11] Mei Y, Sun C, Li C, et al. Research on intelligent assembly method of aero-engine multi-stage rotors based on SVM and variable-step AFSA-BP neural network[J]. Advanced Engineering Informatics, 2021, 54: 101798.

[12] Li L, Chen K, Gao J, et al. Research on optimizing-assembly and optimizing-adjustment technologies of aero-engine fan rotor blades[J]. Advanced Engineering Informatics, 2022, 51: 101506.

[13] Kang H, Li Z. Assembly research of aero-engine casing involving bolted connection based on rigid-compliant coupling assembly deviation modeling[J]. Proceedings of the Institution of Mechanical Engineers, Part C: Journal of Mechanical Engineering Science, 2020, 234(14): 2803-2820.

[14] Tlija M, Korbi A, Louhichi B, et al. A novel model for the tolerancing of nonrigid part assemblies in computer aided design[J]. Journal of Computing and Information Science in Engineering, 2019, 19(4):1-22.

[15] Ballu A, Yan X. Tolerance analysis using skin model shapes and linear complementarity conditions[J]. Journal of Manufacturing Systems, 2018, 48(A): 140-156.

[16] Yi Y, Yan Y, Liu X, et al. Digital twin-based smart assembly process design and application framework for complex products and its case study[J]. Journal of Manufacturing Systems, 2021, 58: 94-107.

[17] Sun X, Bao J, Li J, et al. A digital twin-driven approach for the assembly-commissioning of high precision products[J]. Robotics and Computer-Integrated Manufacturing, 2020, 61: 101839.

[18] Masnaoui W, DaidiE A, Lachaud F, et al. Semi-analytical model development for preliminary study of 3D woven Composite/Metallic flange bolted assemblies[J]. Composite Structures, 2021, 255(1): 112906.

[19] Dembo A, Zeitouni O. Large deviations techniques and applications[M]. New York:

Springer, 2009.

[20] Zhang L, Wang H, Li S, et al. Quantitative modeling and decoupling method for assembly deformation analysis considering residual stress from manufacturing process[J]. Journal of Aerospace Engineering, 2015, 28(3): 04014073.

[21] Corrado A, Polini W. Methods of influence coefficients to evaluate stress and deviation distributions of parts under operating conditions-a review[J]. Engineering Solid Mechanics, 2020, 9(1): 41-54.

[22] Stricher A, Champaney L, Thiebaut F, et al. Tolerance analysis of compliant assemblies using FEM simulations and modal description of shape defects[C]. Nantes: ASME 2012 11th Biennial Conference on Engineering Systems Design and Analysis, 2012.

[23] Zhang Z, Liu J, Anwer N, et al. Integration of surface deformations into polytope-based tolerance analysis: application to an over-constrained mechanism[J]. Procedia Cirp, 2020, 92: 21-26.

[24] Wen Y, Yue X, Hunt J, et al. Virtual assembly and residual stress analysis for the composite fuselage assembly process[J]. Journal of Manufacturing Systems, 2019, 52(7): 55-62.

[25] Bernd G, Maik G, Tobias H, et al. Calculation method for the determination of stress concentrations in fibre-reinforced multilayered composites due to metallic interference-fit bolt[J]. Journal of Composite Materials, 2018, 52(18): 2415-2429.

[26] Wu W, Shang J, Cao Y, et al. Research on applicability of sensitivity table method in optical system alignment[J]. IOP Conference Series: Earth and Environmental Science, 2020, 440(3): 032027.

[27] Corrado A, Polini W. Assembly design in aeronautic field: From assembly jigs to tolerance analysis[J]. Proceedings of the Institution of Mechanical Engineers Part B: Journal of Engineering Manufacture, 2016, 231(14): 2652-2663.

[28] Yoshizato A. Prediction and minimization of excessive distortions and residual stresses in compliant assembled structures[D]. Victoria: University of Victoria, 2020.

[29] Guo F, Liu J, Zou F, et al. Aircraft assembly quality control with feedback actions and assembly station flowing fluctuation analysis[J]. IEEE Access, 2020, 8: 190118-190135.

[30] Smith J. Concept development of an automated shim cell for F-35 forward fuselage outer mold line control[D]. Menomonie: University of Wisconsin-Stout, 2011.

[31] Falgarone H, François T, Coloos J, et al. Variation simulation during assembly of non-rigid components. realistic assembly simulation with ANATOLEFLEX software[J]. Procedia CIRP, 2016, 43: 202-207.

[32] Heling B, Aschenbrenner A, Walter M, et al. On Connected tolerances in statistical tolerance-cost-optimization of assemblies with interrelated dimension chains[J]. Procedia Cirp, 2016, 43: 262-267.

[33] Tlija M, Ghali M, Aifaoui N. Integrated CAD tolerancing model based on difficulty coefficient evaluation and lagrange multiplier[J]. International Journal of Advanced Manufacturing Technology, 2019,101(1): 2519-2532.

[34] Wang H, Lin Y, Yan C. T-Maps-based tolerance analysis of composites assembly involving compensation strategies[J]. Journal of Computing and Information Science in Engineering, 2022, 22(8): 041007.

Digital Assembly Coordination and Quality Controlling Technology
for Aeronautical Thin-walled Structures

[35] Qie Y, Anwer N. Data-driven deviation generation for non-ideal surfaces of Skin Model Shapes[J]. Procedia CIRP, 2022, 109: 1-6.

[36] Zhang J, Qiao L, Huang Z, et al. An approach to analyze the position and orientation between two parts assembled by non-ideal planes[J]. Proceedings of the Institution of Mechanical Engineers, Part B: Journal of Engineering Manufacture, 2021, 235(1-2): 41-53.

[37] Zhang L, Wang H, Li S, et al. Variation propagation modeling and pattern mapping method for aircraft assembly structure considering residual stress from manufacturing process[J]. Proceedings of the Institution of Mechanical Engineers, Part B: Journal of Engineering Manufacture, 2017, 231(3): 437-453.

[38] Chen P. Generalization of Gartner–Ellis theorem[J]. IEEE Transactions on Information Theory, 2000, 46(7): 2752-2760.

[39] Yang J, Cai W, Ran Y, et al. Online measurement based adaptive scalable video transmission in energy harvesting aided wireless systems[J]. IEEE Transactions on Vehicular Technology, 2016, 66(7): 6231-6245.

[40] Wang B, Li H. A sliding window based dynamic spatiotemporal modeling for distributed parameter systems with time-dependent boundary conditions[J]. IEEE Transactions on Industrial Informatics, 2018, 15(4): 2044-2053.

[41] Ratnaweera A, Halgamuge S, Watson H. Self-organizing hierarchical particle swarm optimizer with time-varying acceleration coefficients[J]. IEEE Transactions on Evolutionary Computation, 2004, 8(3): 240-255.

Figure 1-2   Schematic diagram of manufacturing sequence flow with multi-station processes for aircraft

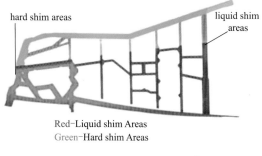

Red−Liquid shim Areas
Green−Hard shim Areas

Figure 1-4   F-35 forward fuselage shimmed surfaces[13]

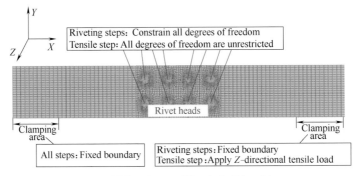

(a) Boundary conditions in the FE model

(b) Von Mises stress of the riveted butt joints under tensile load

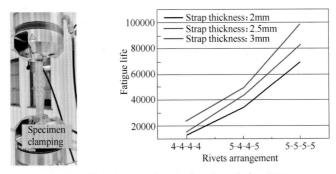

(c) Fatigue experiment and results analysis

Figure 1-6   Investigation on the residual stresses and fatigue performance of riveted single strap butt joints[23]

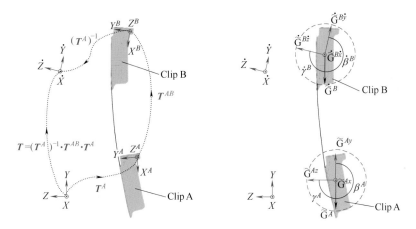

(a) Substructure transformation from Clip A to B

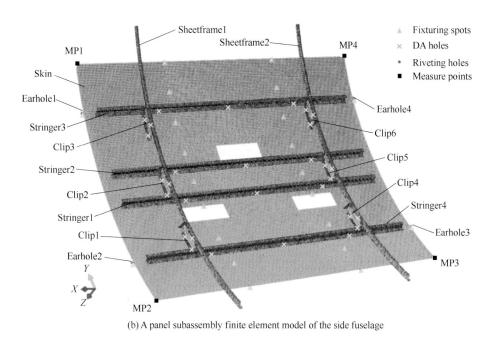

(b) A panel subassembly finite element model of the side fuselage

Figure 1-8　Compliant assembly variation analysis of aeronautical panels with unified substructures[32,33]

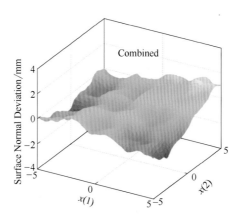

Figure 1-11　Non-ideal part expression with complex distribution model of surface normal deviation[38]

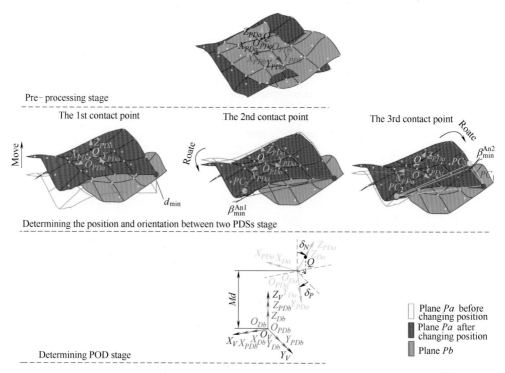

Figure 1-14　Modeling process for PODs (position and orientation deviations)[47]

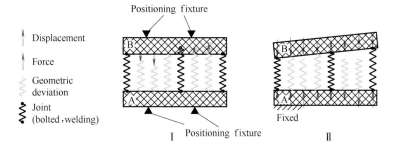

(a) Two types of initial constraint for contact equilibrium status searching

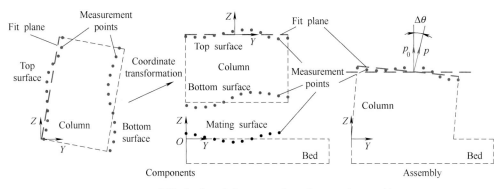

(b) Evaluation of the measured angular error in assembly

Figure 1-17   Assembly variation propagation considering geometric variation and deformation stiffness matrix [56]

Figure 1-27   Automated joining and flexible holding fixtures for CFRP panel[74]

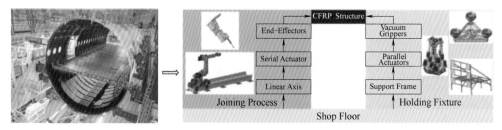

Figure 1-28 Flexible Automated Assembly System (FAAS) for CFRP-panels of the Airbus A350XWB[74]

Figure 1-32 The CHARLIE integration verification component in StarDrive plan[3]

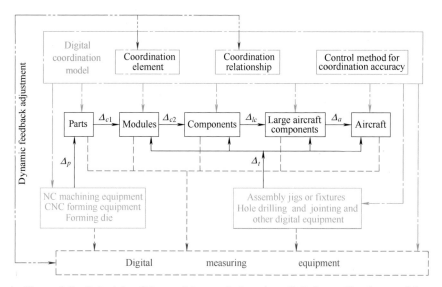

Figure 2-3 Principle of the working mode based on digital coordination model

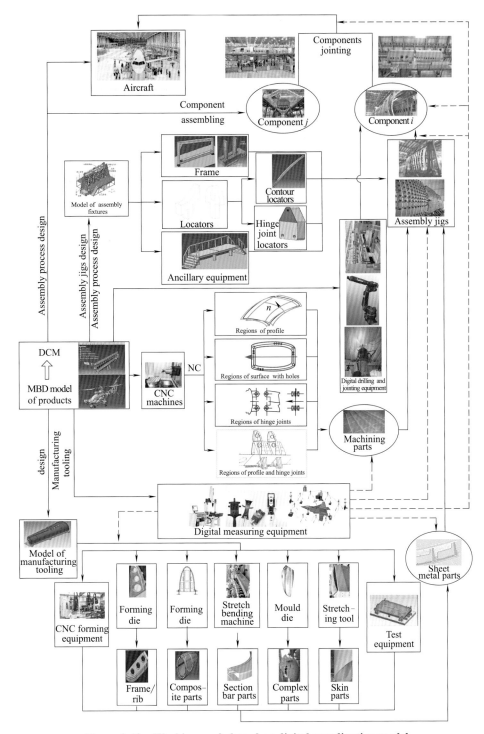

Figure 2-12　Working mode based on digital coordination model

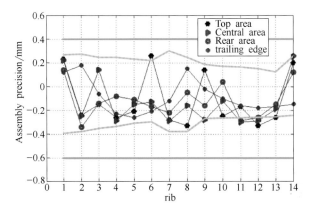

Figure 2-22　Comparison of skin profile precision for the inner component

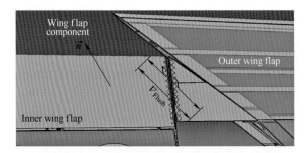

Figure 3-9　Flush requirement between different components

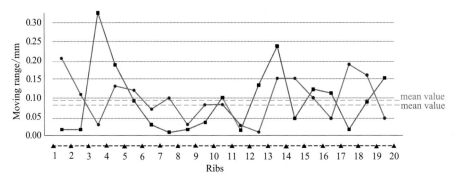

Figure 3-13　Moving range of the precision data

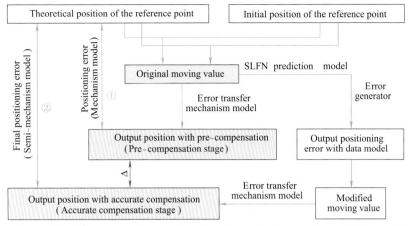

Figure 4-15　Comprehensive combination of mechanism model and data model

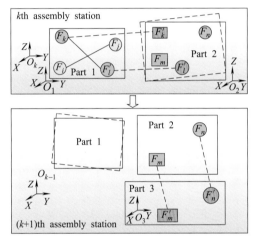

Figure 5-8　Generating principle of relocation error

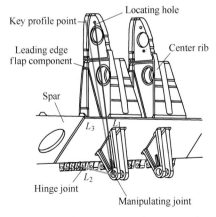

Figure 5-12　Transformation of the coordination datum in the jointing process

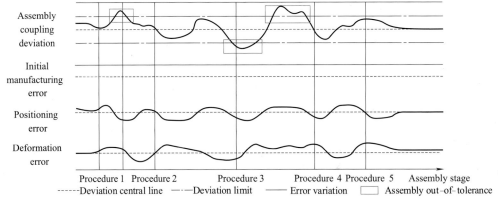

Assembly coupling deviation

Initial manufacturing error

Positioning error

Deformation error

Procedure 1　Procedure 2　　　　Procedure 3　　　Procedure 4　Procedure 5　　Assembly stage

------Deviation central line　—·—Deviation limit　——Error variation　☐ Assembly out-of-tolerance

Figure 6-2　Assembly error generating and coupling at different assembly stages

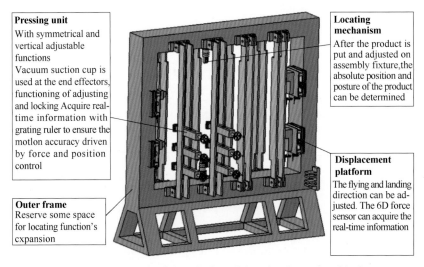

Pressing unit
With symmetrical and vertical adjustable functions
Vacuum suction cup is used at the end effectors, functioning of adjusting and locking Acquire real-time information with grating ruler to ensure the motlon accuracy driven by force and position control

Locating mechanism
After the product is put and adjusted on assembly fixture,the absolute position and posture of the product can be determined

Displacement platform
The flying and landing direction can be adjusted. The 6D force sensor can acquire the real-time information

Outer frame
Reserve some space for locating function's cxpansion

Figure 6-9　Structure and description of the wing-box assembly fixture